LITERATURE AND SCIENCE AS MODES OF EXPRESSION

BOSTON STUDIES IN THE PHILOSOPHY OF SCIENCE

Editor

ROBERT S. COHEN, *Boston University*

Editorial Advisory Board

ADOLF GRÜNBAUM, *University of Pittsburgh*
SYLVAN S. SCHWEBER, *Brandeis University*
JOHN J. STACHEL, *Boston University*
MARX W. WARTOFSKY, *Baruch College of the City University of New York*

VOLUME 115

PENSACOLA JR. COLLEGE LRC

Logo of the Society for Literature and Science prepared by Dr Lance Schachterle.

LITERATURE AND SCIENCE AS MODES OF EXPRESSION

Edited by
FREDERICK AMRINE

with an Introduction by Stephen J. Weininger

KLUWER ACADEMIC PUBLISHERS
DORDRECHT / BOSTON / LONDON

Library of Congress Cataloging-in-Publication Data

```
Literature & science as modes of expression / edited by Frederick
  Amrine and Robert S. Cohen ; with an introduction by Stephen J.
  Weininger.
       p.   cm. -- (Boston studies in the philosophy of science ; v.
  115)
     Includes index.
     ISBN 0-7923-0133-1
     1. Literature and science.  2. Science in literature.
  3. Literature and technology.   I. Amrine, Frederick, 1952-    .
  II. Cohen, Robert Sonné.  III. Title: Literature and science as
  modes of expression.  IV. Series.
  Q174.B67  vol. 115
  [PN55]
  001'.01 s--dc19
  [801]                                                       89-2812
```

Published by Kluwer Academic Publishers,
P.O. Box 17, 3300 AA Dordrecht, The Netherlands.

Kluwer Academic Publishers incorporates the publishing programmes of
D. Reidel, Martinus Nijhoff, Dr W. Junk and MTP Press.

Sold and distributed in the U.S.A. and Canada
by Kluwer Academic Publishers,
101 Philip Drive, Norwell, MA 02061, U.S.A.

In all other countries, sold and distributed
by Kluwer Academic Publishers Group,
P.O. Box 322, 3300 AH Dordrecht, The Netherlands.

All Rights Reserved
© 1989 Kluwer Academic Publishers
No part of the material protected by this copyright notice may be reproduced or
utilized in any form or by any means, electronic or mechanical,
including photocopying, recording or by any information storage and
retrieval system, without written permission from the copyright owner.

Printed in the Netherlands

LITERATURE AND SCIENCE AS MODES OF EXPRESSION

Edited by
FREDERICK AMRINE

with an Introduction by Stephen J. Weininger

KLUWER ACADEMIC PUBLISHERS
DORDRECHT / BOSTON / LONDON

TABLE OF CONTENTS

EDITORIAL PREFACE	ix
INTRODUCTION: THE EVOLUTION OF LITERATURE AND SCIENCE AS A DISCIPLINE	xiii
GILLIAN BEER / Discourses of the Island	1
G. S. ROUSSEAU / Discourses of the Nerve	29
WALTER MOSER / Experiment and Fiction	61
ROBERT KOCH / Hypotyposes	81
KENNETH J. KNOESPEL / The Mythological Transformations of Renaissance Science: Physical Allegory and the Crisis of Alchemical Narrative	99
GÉZA VON MOLNÁR / "What Ever Happened to Ethics?"	113
JOHN NEUBAUER / Nature as Construct	129
PAIGE MATTHEY BYNUM / "Observe how healthily — how calmly I can tell you the whole story": Moral Insanity and Edgar Allan Poe's 'The Tell-Tale Heart'	141
JOSEPH W. SLADE / Conceptualizing Technology in Literary Terms: Some American Examples	153
MILES ORVELL / Literature and the Authority of Technology	169
STEVEN CARTER / "A Place to Step Further": Jack Spicer's Quantum Poetics	177
INDEX OF NAMES	189

literary and linguistic theory, ethics, and the histories of literature, medicine, psychiatry, and technology. They demonstrate that such interdisciplinary inquiry can be pursued rigorously, and is capable of yielding unique insights. We are convinced that the most significant gains in both the history and philosophy of science will be accomplished henceforth by scholars working along such lines. What is needed now is not so much another treatise on "The Nerve in the History of Science," nor on "The Literature of Sensibility"; not so much on "The Theme of the Island in Literature", nor on "Scientific Studies of Island Ecology", but rather more studies like our exemplary plenary talks by Gillian Beer and George Rousseau on "Discourses of the Island" and "Discourses of the Nerve". As George Levine has written, "science and literature reflect each other because they draw mutually on one culture, from the same source, and they work out in different languages the same project" (p. 7). Pursuit of insular disciplines needs now to be complemented by broader "cultural studies" such as these.

In his introduction, Weininger rightly emphasizes the advances that have been made by the field as a whole and our contributors in particular. Nevertheless, it seems to us that several important lines of inquiry taken up in the present volume deserve further attention. One fundamental question has to do with the epistemic value of metaphor and literature generally: good work on the former has been done by Mary Hesse among others, but, as Rousseau has argued, the latter remains largely "an unexplored territory" (p. 587). And in what ways does what Weininger calls the "influence vector" point from science toward literature? We have gotten beyond seeing literature as merely the passive receptacle for scientific 'themes', but are there other important ways in which science influences — or could influence — literature? Are there e.g. important lessons that the literary theorist might learn from the philosopher of science? It may be that "naive" realism has been routed from the philosophy of science once and for all, but does this mean that we must run to the other extreme (as Livingston complains has too often been the case), arguing that science is but another form of fiction — or even an elaborate myth? Is there some middle ground to be staked out between the extremes of positivist reduction and total relativism? Our hope is that *Literature and Science as Modes of Expression* will promote further debate on these important questions.

Finally, we would like to thank the following colleagues who con-

EDITORIAL PREFACE

On the 25th anniversary of the founding of the Boston Studies series in 1985, Cohen, Elkana, and Wartofsky wrote in another preface such as this that the time had come for establishing institutions supporting a vision to which the series had been devoted since its inception, namely that of a more broadly conceived, interdisciplinary study of the history and philosophy of science:

> In recent years it has become evident that, in addition to serious and competent disciplinary work on the specifics of the History of Science, the Philosophy of Science and the Sociology of Science, there is now a growing need to develop a problem-oriented approach which no longer distinguishes between these three specialties in a cut and dried way. Since the time has come for such an approach, the institutional tools should be provided.
>
> A way to do so would be ... to organize colloquia and to publish good papers stemming from these, without attempting to organize the papers under the separate rubrics of History of Philosophy or Sociology of Science; and moreover to consider it natural that any fundamental issue of the foundations of the sciences, or their place in a culture and the way they are institutionalized in the societal web, is still our concern, no matter whether we are a professional scientist, historian or philosopher who deals with the problem (p. vii).

In retrospect, their plea reveals itself to have been remarkably prophetic, for it was in the same year that the Society for Literature and Science was inaugurated in order to foster the same vision. Hence it seems to us altogether fitting that papers from the first annual conference of the Society should now appear in the Boston Studies series. Although the study of literature and science is hardly new (see Stephen J. Weininger's elegant introduction for a capsule history), we feel that the essays collected here constitute both the record of a historic occasion and a testament to the maturation of the field.

If anything, the present volume seeks to extend the fundamental interdisciplinary thrust of the Boston Studies series even further. For the name "Society for Literature and Science" is actually a misnomer. It is much too narrow to encompass the range of "interdisciplinarities" that are pursued beneath its aegis. The essays collected here combine not only the history, philosophy, and sociology of science, but also

Frederick Amrine (ed.), Literature and Science as Modes of Expression, ix–xi.

tributed greatly to the preparation of this volume by serving as its editorial board:

Mordechi Feingold (Boston University)
Sander Gilman (Cornell University)
N. Katherine Hayles (University of Iowa)
George Levine (Rutgers)
James Paradis (Massachusetts Institute of Technology)
Lissa Roberts (Portland State University)
Arthur Quinn (University of California, Berkeley)
Stephen J. Weininger (Worcester Polytechnic Institute)

REFERENCES

Cohen, R., Y. Elkana and M. Wartofsky: 'Editorial Note', in *The Kaleidoscope of Science: The Israel Colloquium: Studies in History, Philosophy, and Sociology of Science, Vol. 1*, ed. E. Margalit, Boston Studies in the Philosophy of Science, Vol. 94, D. Reidel, Dordrecht, 1985, pp. vii–viii.

Levine, G., ed.: *One Culture*, Essays in Science and Literature, Univ. of Wisconsin Press, Madison, 1987.

Livingston, P.: *Literary Knowledge: Humanistic Inquiry and the Philosophy of Science*, Cornell Univ. Press, Ithaca and London, 1988.

Rousseau, G.: 'Literature and Science: The State of the Field', *ISIS* **69** (1978) pp. 583—91.

The University of Michigan   FREDERICK AMRINE
Boston University   ROBERT S. COHEN

something in the conjunction that seems inherently problematic, because of the perceived role of science in society.

Concurrent with the acceptance of science as a part of culture came the conviction that science also possessed a unique cultural autonomy: no other product of Western history seemed so ahistorical and thus immune to the vicissitudes of cultural change. This conclusion in turn enforced the perception of a highly asymmetric relation between science and the rest of culture. Science seemed always to be the active component, influencing some domain of essentially passive culture.

As a result, Literature and Science as an academic activity began as a study of the incorporation of scientific ideas into literary texts. Incorporation took several forms — scientific theories could directly alter the writers' perception of the world, or supply metaphors by which to capture it. The task of the cultural historian was then to ferret out the traces of these influences. One of the distinguished practitioners of this sort of cultural history was Marjorie Hope Nicolson, whose *Newton Demands the Muse* exemplifies the value of Literature and Science studies for the History of Ideas.

The complementary activity, seeking literary influences in scientific texts, was relatively rare unless the texts were quite old. Yet a passage such as this:

As, in the imagination of Dante, the invisible air becomes peopled with spiritual beings, so before the eyes of earnest investigators, and especially before the eyes of Clerk Maxwell, the invisible mass of gases became peopled with particles... (Mendeléef)

suggests that the language of more contemporary scientists might also have borne examination.

A sea change in the orientation of literature and science studies began in the 1960s, induced in large measure by the new directions in which the history and philosophy of science were moving (Neubauer, 1983b), and by the (in)famous *Two Cultures* debate (Schachterle). The latter issue inspired T. H. Huxley's grandson Aldous to publish his essay *Literature and Science*. From a contemporary perspective the younger Huxley's essay is intriguingly Janus-like: it adumbrates several important lines of future development while remaining captive to some very conventional notions of the relationship between literature and science.

Huxley was very firm in his conviction that science is a powerful and

INTRODUCTION: THE EVOLUTION OF LITERATURE AND SCIENCE AS A DISCIPLINE

In August, 1985 the Society for Literature and Science was officially inaugurated at the International Congress for the History of Science, held at the University of California at Berkeley (Rousseau, 1986). Two years later the Society held its first convention, 'Literature and Science as Modes of Expression,' in Worcester, Massachusetts. The program confirmed that the field was alive with activity — there were no fewer than three plenary lectures, and about 75 contributed papers grouped into 19 sessions. The readers of this volume will be sampling two of the plenary lectures and 10 of the papers.

Given that academic societies and conferences come and go, it would not be unreasonable to ask why this pair merit particular attention. It is not the case that a new field came into being — the relations of literature and science have been the object of study for some time. Already at the end of the nineteenth century T. H. Huxley's lecture on 'Science and Culture' provoked Matthew Arnold's famous reply, 'Literature and Science' (Cadden and Brostowin). The subject has a long and distinguished ancestry. But its objects, aims and outlooks have undergone some dramatic changes over time, and perhaps never more so than in recent years. The present state and future prospects for Literature and Science were one of the major themes of the Worcester conference and of this volume.

The Huxley-Arnold exchange which initially defined the field concerned itself, among other things, with what constituted culture and which of its components were worth preserving and transmitting. By the twentieth century most educated people, whether reluctantly or enthusiastically, came to accept the proposition that science was an integral part of Western culture. A logical consequence of this conclusion was that the cultural influence of science is a valid and important area of intellectual inquiry, an idea that gained ground slowly and arduously among scholars. The obstacles to establishing the cultural relations of science as a legitimate academic topic arose, in part, from institutional pressures and conflicts that seem only too predictable and familiar. Yet academic politics alone will not account for the difficulties; there is

Frederick Amrine (ed.) Literature and Science as Modes of Expression, xiii–xxv.
© *1989 Kluwer Academic Publishers, all rights reserved.*

overview of the field. Moreover, the changes in literary theory were paralleled by new directions in the history and philosophy of science, which made it highly improbable that the traditional view of scientific language — rhetorically neutral, figuratively unmarked, inherently different from literary language — could continue to prevail (Carlisle; Neubauer, 1983b; Schuster and Yeo). And as this view went into eclipse it carried many traditional approaches to literature and science into the shadows as well. A reconceptualization of the field was clearly in order.

Thus, studies of the *scientific* role of metaphor and of the relations of models and metaphors signaled an important new direction (Black; Hesse). Carlisle's paper drew attention to the ways in which semiotic and discourse analyses further undermined the distinctions between literary and scientific language. It even became possible to debate whether understanding science entails anything *more than* analyzing discourse (Mulkay; Shapin; Woolgar) — a startling measure of how much things had changed in a relatively short period. As theoretical perspectives multiplied for Literature and Science, its newer contributions reflected the field's expanding horizons. Beer uncovered Darwin's metaphors and rhetorical procedures; Hayles explored the parallels between "scientific field models and literary strategies in the twentieth century"; and Weininger examined the problem of the shifting meaning of scientific terms as a function of their explanatory context.

Rousseau's 1978 article was the first of several assessments of the state of the field in the last decade (Rousseau, 1981; Guffey and Slusser; Neubauer, 1983a, b, 1987; Peterfreund). When a patient's temperature is taken at such regular intervals it is reasonable to infer some feverish activity, a diagnosis lent further credibility by the rash of conferences devoted to Literature and Science after 1978: in Luxembourg, sponsored by Clark University, and at the Library of Congress (*Science and Literature*), both in 1981; at Long Island University in 1983 (Slade and Lee); in Worcester, sponsored by the Society for Literature and Science, in 1987; at the Society for the Humanities at Cornell University for 1987—88; and at the Twentieth-Century Literature Conference at the University of Louisville in 1988. Further evidence that the field is flourishing may be gleaned from the recent appearance of essay collections devoted to it (Jordanova; Rousseau, 1987), one of which (Levine) is the first of a series. A centennial bibliography covering the years 1880—1980, produced

pervasive aspect of everyday life that demands the attention of writers and poets, but to which they had responded quite inadequately. He also saw that the epistemological thrust of modern science was eroding traditional distinctions between subject and object, inner and outer, and also eroding the boundaries between the worlds of the scientist and the artist. His claims that modern science and technology offer the writer a cornucopia of "raw material" for the imagination, and that artists and scientists share wellsprings of creativity, have remained convincing (Woodcock).

Huxley was less prescient in his insistence on the diametrical opposition between literary and scientific language. According to him, common language is equally inadequate for the needs of scientists and writers, but that is their only point of agreement where language is concerned.

In the scientist verbal caution ranks among the highest of virtues. His words must have a one-to-one relationship with some specific class of data or sequence of ideas. By the rules of the scientific game he is forbidden to say more than one thing at a time ... (p. 36)

By contrast,

Poets and, in general, men of letters, are permitted, indeed even commanded, by the rules of *their* game, to do all the things that scientists are not permitted to do. (pp. 36—7)

For Huxley language is a great chasm that separates science from literature; on one side parsimony and discipline, on the other effulgence and abandon.

That a thinker of Huxley's acumen could accept the alleged polarity between scientific and literary prose in so simplified a form testifies to the power of the 'official' view of scientific language, promulgated and believed by scientists and non-scientists alike. A reassessment of that dogma was inevitable, however. The continuing examination of the manifold social relations of science led inexorably to an inquiry into the linguistic foundations of science.

The motivation and methodology for this inquiry came partly from literary criticism and philosophy, particularly from the Continent. The significance of these developments for the future prospects of Literature and Science was pointed out by G. S. Rousseau, in his 1978

interchange across disciplinary lines, and extensive intermingling of tropes and metaphors. Demonstrating these interactions usually entails either a comparison of contemporaneous scientific and literary texts, or a study of the parallel evolution of the two discourses over time. Whatever the method, some of the most striking cases are to be found among the human sciences (Jordanova). For example, Paige Bynum shows how attentive Poe was to contemporaneous theories of insanity, which are central to his story, "The Tell-Tale Heart". These scientific-medical theories raised very troubling questions about the human capacity for violent, asocial behavior, concerns that accounted for the story's initial warm reception and for its continuing power, even though the science on which it was based has long been discarded.

G. S. Rousseau's essay on "Discourses of the Nerve" is also located in the sphere of the human sciences, set in a time when the discourses of literature and medicine were anything but autonomous. There are several striking features of this period that are noteworthy. Rousseau suggests that talk about sensibility may have been imported *into* science *from* literature, a reverse direction for the 'influence' vector as usually conceived. More significantly, he shows that by treating both disciplines as discourses, we can the more readily recognize not only their common social roots, but also the specific social interests — class, gender — that they reflect. Here then is one of the important aims of Literature and Science: to investigate the social roots and social allegiances of both discourses in tandem, the better to place each in its context.

While literary entanglements may lie closer to the surface in the human compared to the physical sciences, they are just as surely there in the latter. The history of chemistry is, in fact, a paradigm case for the centrality of language to scientific development. Two of its most crucial transformations were accompanied by struggles over language that aimed, successfully, at instituting new didactic practices, and which were as consequential as any change in laboratory methodology (Hannaway; Anderson). Kenneth Knoespel enriches our understanding of this process by describing how narrative description, supposedly banned from chemical discourse after it replaced alchemical discourse, went 'underground' by being condensed into powerful metaphors that persisted into the eighteenth century.

The period covered by Knoespel and Rousseau was 'prelapsarian', in the sense that the unbridled growth of differentiated discourses had not yet taken place, while Gillian Beer writes about a distinctly 'postlapsarian' era. Her theme is 'the island', a topic that recurs in

under the auspices of the Modern Language Association, has also appeared (Schatzberg *et al.*).

What, then, has been the effect of all this ferment on the range of disciplinary sources and methods? How should the field of literature and science be characterized with regard to its aims and its products? Does it have boundaries that demarcate it from the history of science and literary history, from the philosophy and sociology of science? The essays in this volume propose answers to at least some of these questions, and they raise issues which point toward others.

Works of fiction have traditionally provided the bulk of the texts for literature and science studies, and they are well represented in this volume: the medieval romance (Knoespel), novels and stories by Poe (Bynum) and Dos Passos (Orvell), the poetry of Matthew Arnold (Beer), T. S. Eliot (Slade) and Jack Spicer (Carter). Literary writers, such as Hart Crane (Slade) and James Agee (Orvell), are also represented by their nonfiction works. A newer trend revolves around the literary scrutiny of scientific texts; that trend is exemplified here by Beer's examination of Wegener and Darwin, alongside the writings of other scientific and literary notables.

Moreover, our pool of sources is fed by more than literary and scientific streams. It encompasses the writings of philosophers such as Kant (Koch), Condillac and Mach (Moser), and of the polymathic philosopher/poet/geologist Novalis (Neubauer and von Molnár). Novalis is a unique figure. His attempts toward a synthesis of diverse discourses was possible, in part, because he worked at a time when it was still possible to envision a 'poeticized science'.

We are thus reminded that the problematic that engendered this volume — the relations between literature and science — is of relatively recent origin, no earlier than roughly mid-eighteenth century. "Was the Bernard Mandeville who wrote *The Fable of the Bees* [1705] as well as the treatise on hysteria primarily a doctor or a writer?" asks Rousseau in his essay. The next sentence brings the appropriate answer: "The question is as bad as the false dichotomy ...". Thus one of the major focuses for Literature and Science is the process whereby two separate discourses were formed out of one (Bazerman), since from the our current perspective such a separation does not seem ineluctable (Peterfreund, p. 30).

Not only was the separation not inevitable, it was far from complete. All the authors assembled here are unanimous in seeing extensive

Walter Moser shows how commonplace they were in the eighteenth century, particularly in discussions of the origin of one or another human institution. By the end of that century, however, an increasing distrust of hypotheses and fictions was apparent, especially in the sciences. Establishing a clear demarcation between fact and fiction was a matter of urgency in the nineteenth century. By the time Mach wrote *Erkenntnis und Irrtum* in 1905, the difference was unambiguous. Thus, when he called the *Gedankenexperiment* the common ancestor of both fiction and the true scientific experiment, he was able to make clear exactly how they had diverged. Unlike the work of fiction, the scientific experiment "has to respect the given structure of the factual world and correspond to it . . .". But how sharply are *we* to take the distinction if we acknowledge that in some way the "given structure of the factual world" is constructed by the very same language which constructs the worlds of fiction?

Part of the difficulty we have in sorting out the relation between language and science arises from our habit of thinking of language as representation *independent of action* (Keller). Doing so confronts us inevitably with sets of opposed concepts such as 'language' and 'method', or 'theory' and 'practice', or 'representing' and 'intervening' (Hacking). But here Novalis has something to say that I find very helpful. Novalis conceives of language as a bridge between theoretical reason (science) and practical reason (ethics). He is able to do so, according to Géza von Molnár, because he sees science, ethics, and language as sharing two essential characteristics: they are each oriented toward action in the world, and they are each realized in a social context. Furthermore, the interrelations of the three fields entail mutual modification.

In such a framework it makes sense to talk, as Steven Carter does, about Jack Spicer's attempt to erase the "ideas and structures which are imposed upon language by Newtonian epistemological systems", in order to "present the reader with linguistic models which are relevant to a quantum not a Newtonian universe". Here, then, I believe, is a new and rewarding approach for relating science and language: asking how each of them operates to bring about changes in the world, in a manner and to an extent that depends on a context that invariably includes the other. The essays in this volume help to set us on that path.

Lastly, one must admit that the pursuit of Literature and Science now entails more than the desire to expropriate some new academic

numerous distinct discourses as diverse as geology and psychology. As she disentangles the manifold metaphoric possibilities of this rich *topos*, we are reminded that by no means all the potential metaphoric uses of any term are activated simultaneously. Beer shows us certain connotations attaching themselves to the word 'island' across a range of more or less contemporaneous discourses, and then allows us to watch these change as the social context changes. Her exposition leads us toward the inescapable question that Rousseau, in the second part of his essay, poses for all practitioners of the discipline: "What is your *new* theory of Literature and Science?" It is unavoidable because our hopes of assimilating new theoretical perspectives from literary criticism and history and philosophy of science, and of charting new directions, cannot be successful in the absence of a theory of our own field.

Since the decline of the 'influence' model, this has been a difficult question indeed. In her answer, Beer tries to steer between the Scylla of simplistic causality and the Charybdis of random happenstance:

> ... it would be forcing the evidence, I believe, to propose a *causal* relationship in either direction between literary and scientific theory. Neither, though, is a *casual* relationship entirely convincing. ... We cannot *derive* literary from scientific revolutions directly: postmodernism is collateral with plate tectonics, not dependent on it. Nor can we *depend* on literature to provide blueprints or even harbingers of future scientific developments. The relationship between these diverse human activities is deeper and more mobile. Both draw on, are to some degree controlled by, and in their turn help to form, the common anxieties of the time.

If we want to explore this deeper relationship, then we must come to grips with the specific role of language. In his discussion of Kant's *hypotyposes*, Robert Koch asserts that since language "in fact constructs the very worlds it purports to describe", there is no "unbridgeable abyss" between science and literature; they "can be thought according to their identity *in* language, their differences being ... merely modes of expression". And in the writings of Novalis, John Neubauer finds a strongly constructivist theory of science, as well as a theory of poetics grounded in scientific metaphors. Novalis' insight into the centrality of the imagination for all human constructions, including science, led him to advocate "experimentation with images and concepts in the imagination", of a kind which Neubauer finds very close to what we know as thought experiments. Now, thinking about thought experiments can help us clarify our thinking about the relations between language and science.

Thought experiments, or their equivalents, have a venerable ancestry.

problematic. For instance, it is necessary to ask about the status of scientific writings not intended for the specialized, subdisciplinary readership. Are they part of 'science'? If so, then the distinction between the audiences for scientific and nonscientific texts loses some of its force. Analogously, the choice of what is to count as 'literature' has a profound effect on the conclusions we reach about the compatibility of literature and science. At one point Kipperman's article counsels caution with respect to a theory "of science and *belles lettres*" (p. 77). Is literature coextensive with *belles lettres*?

Here I would draw attention to the essays in this volume by Miles Orvell and Joseph Slade on the relations of literature and *technology*. Both make the case, indirectly but effectively, for an expansion of the range of texts usually subsumed under the rubrics 'science' and 'literature'. As Slade points out, the recognition that the essence of technology was not objects but knowledge has profoundly affected the writer's relation to, and vision of, technology. The tradition of grouping literature and science together as intellectual pursuits, in opposition to technology, has became less and less tenable. Orvell's quote from Dos Passos is to the point: "In his relation to society a professional writer is a technician just as much as an electrical engineer is". Writing itself is a technology, and its dissemination is a complex and pervasive technology. If we accept this then we will realize that in many ways science, technology, and literature overlap along a continuum. What, then, is the effect of this recognition on our sense of purpose?

At the conclusion of his essay in this volume Rousseau notes that one function for Literature and Science may be "science criticism"; Levine argues that "it has opened the possibility of a serious critique of science that is not merely sentimental or alarmist" (p. 18). I agree, but would add a codicil: it must also produce a new understanding of the nature of *literature* and of *language* (Shaffer), and ultimately of our culture. Many of those who resist attempts to bring the humanities and the sciences together decry the latter for its 'coldness' and 'remoteness from everyday experience.' I would contend that what discomfits many of them is not the remoteness of science but rather its proximity. To acknowledge its nearness is to risk undermining a cherished and comforting image of 'culture'. The price of maintaining that myth is too high for all of us, regardless of our disciplinary loyalties.

Facilitating the presentation and publication of these papers has been a great privilege for me. I am truly grateful to all my colleagues

Lebensraum; there is a commitment, explicit or otherwise, to what George Levine has identified as *One Culture*. It is a view that stresses the similarities between the sciences and the humanities, and downplays their differences. While this view has been increasingly favored over the last several decades, it does not command universal assent. Thus it behooves us to pay attention to those who see more difference than similarity between science and literature.

These critics often make their point by focusing on the *audiences* for the two discourses. If we restrict our attention to the primary texts of each discourse, then there is indeed a distinct difference in attitude with respect to the size of the potential audience. Markus has made a detailed comparison of the relationship between author, text, and reader in scientific and non-scientific discourses. He affirms the oft-remarked exclusivity of scientific texts, intended as they are for a narrow band of expert readers. However, the important point is that this exclusivity is *not* an inevitable consequence of 'technical jargon' or 'specialization'. It is rather part of the structure of scientific discourse and serves distinct functions. Gillian Beer (1987) believes that one of these functions is to control the inherent tendency of language to proliferate meanings. By limiting audience size one can, she proposes, impose a tacit agreement on readers as to the range of allowable meanings, at least in the short term.

There is also the matter of the intended *effect* on the audience. Steinberg has recently argued that the reception accorded works of art is so different from that accorded scientific works as to overwhelm any similarities. A related argument has recently been deployed specifically against attempts to establish a theory of Literature and Science; we are warned that their rhetorical aims are so incompatible "that science and literature not only cannot be 'reconciled' but also they must not be" (Kipperman, p. 76). It seems clear then that questions about the audiences for the two discourses and the ways in which they are received are critical for any theory of Literature and Science. These questions deserve very searching and detailed answers; to my knowledge, they have not been forthcoming.

Yet I cannot refrain from taking issue with Steinberg and Kipperman, particularly with the stark manner in which they have posed the separation between the sciences and the humanities. In order to maintain that sharp separation they cricumscribe, mostly by implication, the meanings of 'science', 'art', and 'literature' in ways that I find

Natural Philosophy', in *The Principles of Chemistry*, 2nd edn. (ed. by T. Lawson, trans. by G. Kamensky), Longmans, Green, London, 1987, vol. 2, pp. 453—70. An address at the Royal Institution, 1889.

Mulkay, M.: 'Action and Belief or Scientific Discourse? A possible way of ending intellectual vassalage in social studies of science', *Philosophy of Social Science* **11** (1981) pp. 163—71.

Neubauer, J.: 'Literature and Science: Future Possibilities', *University of Hartford Studies in Literature* **19** (1987) pp. 53—7.

Neubauer, J.: 'Literature and Science: Their Metaphors and Metamorphoses', *Yearbook of Comparative and General Literature* **32** (1983b) pp. 67—75.

Neubauer, J.: 'Models for the History of Literature and Science', in *Science and Literature* (ed. by H. Garvin and J. Heath), Associated Univ. Presses, London and Toronto, 1983a, pp. 17—37.

Nicolson, M.: *Newton Demands the Muse: Newton's Opticks and the Eighteenth Century Poets*, Princeton Univ. Press, Princeton, 1946.

Peterfreund, S.: 'Literature and Science: The Present State of the Field', *University of Hartford Studies in Literature* **19** (1987) pp. 25—36.

Rousseau, G.: 'The Discourse(s) of Literature and Science', *University of Hartford Studies in Literature* **19** (1987) pp. 1—24. This number of the journal is devoted entirely to literature and science.

Rousseau, G.: 'Literature and Medicine: The State of the Field', *Isis* **72** (1981) pp. 406—24.

Rousseau, G.: 'Literature and Science: The State of the Field', *Isis* **69** (1978) pp. 583—91.

Rousseau, G.: 'Science and the Imagination', *Annals of Scholarship* **4** (1986). This entire issue of the journal is devoted to literature and science.

Schachterle, L.: 'What Really Distinguishes the "Two Cultures"?', *Annals of Scholarship* **4** (1986) pp. 83—94.

Schatzberg, W., Waite, R. and Johnson, J., eds.: *The Relations of Literature & Science: An Annotated Bibliography of Scholarship, 1880—1980*, Modern Language Association, New York, 1987.

Schuster, J. and Yeo, R., eds.: *The Politics and Rhetoric of Scientific Method: Historical Studies*, D. Reidel, Dordrecht, 1986.

Science and Literature, Library of Congress, Washington, D.C., 1985.

Shaffer, E.: 'Literature and Science: Toward a New Literary History', *University of Hartford Studies in Literature* **19** (1987) pp. 37—52.

Shapin, S.: 'Talking History: Reflections on Discourse Analysis', *Isis* **75** (1984) pp. 125—30.

Slade, J. and Lee, J. eds.: *Beyond the Two Cultures: Essays in Science, Technology, and Literature*, Iowa State Univ. Press, Ames, forthcoming 1989.

Steinberg, L.: 'Art and Science: Do They Need to be Yoked?' *Daedalus* **115**(3) (1986) 1—15.

Weininger, S.: 'Concept and Context in Contemporary Chemistry', in *Beyond the Two Cultures: Essays in Science, Technology, and Literature* (ed. by J. Slade and J. Lee), Iowa State Univ. Press, Ames, forthcoming 1989.

who shared so much of the effort, but must single out two by name: Lance Schachterle, who as President of SLS did so much to make its inaugural conference a success, and Fred Amrine, who as editor of this volume did so much to bring its fruits to a wider audience.

REFERENCES

Anderson, W.: *Between the Library and the Laboratory*, Johns Hopkins Univ. Press, Baltimore, 1984.
Bazerman, C.: *Shaping Written Knowledge: The Genre and Activity of the Experimental Report*, Univ. of Wisconsin Press, Madiscon, 1988.
Beer, G.: *Darwin's Plots: Evolutionary Narrative in Darwin, George Eliot, and Nineteenth Century Fiction*, Routledge and Kegan Paul, London, 1983.
Beer, G.: 'Problems of Description in the Language of Discovery', in *One Culture: Essays in Science and Literature* (ed. by G. Levine), Univ. of Wisconsin Press, Madison, 1987, pp. 35—58.
Black, M.: *Models and Metaphors: Studies in Language and Philosophy*, Cornell Univ. Press, Ithaca, 1962.
Cadden, J. and Brostowin, P.: *Science and Literature: A Reader*, D. C. Heath, Boston, 1964.
Carlisle, E.: 'Literature, Science, and Language: A Study of Similarity and Difference', *Pre/text* **1** (1980) 39—72.
Guffey, G. and Slusser, G.: 'Literature and Science', in *Interrelations of Literature* (ed. by J.-P. Barricelli and J. Gibaldi), Modern Language Association, New York, 1982, pp. 176—204.
Hacking, I.: *Representing and Intervening: Introductory Topics in the Philosophy of Natural Science*, Cambridge Univ. Press, Cambridge, 1983.
Hannaway, O.: *The Chemists and the World*, Johns Hopkins Univ. Press, Baltimore, 1975.
Hayles, N.: *The Cosmic Web: Scientific Field Models and Literary Strategies in the Twentieth Century*, Cornell Univ. Press, Ithaca and London, 1984.
Hesse, M.: 'The Explanatory Function of Metaphor', in *Logic, Methodology and Philosophy of Science* (ed. by Y. Bar-Hillel), North-Holland, Amsterdam, 1965.
Huxley, A.: *Literature and Science*, Harper and Row, New York, 1963.
Jordanova, L., ed.: *Languages of Nature: Critical Essays on Science and Literature*, Rutgers Univ. Press, New Brunswick, 1986.
Keller, E.: 'Critical Silences in Scientific Discourse: Problems of Form and Re-Form', lecture at the Institute for Advanced Study, Princeton, NJ, 4 Feb. 1988.
Kipperman, M.: 'The Rhetorical Case Against a Theory of Literature and Science', *Philosophy and Literature* **10** (1986) pp. 76—83.
Levine, G., ed.: *One Culture*, Essays in Science and Literature, Univ. of Wisconsin Press, Madison, 1987.
Markus, G.: 'Why Is There No Hermeneutics of Natural Sciences? Some Preliminary Theses', *Science in Context* **1** (1987) pp. 5—51.
Mendeléef, D.: 'An Attempt to Apply to Chemistry One of the Principles of Newton's

Woodcock, J.: 'Literature and Science Since Huxley', *Interdisciplinary Science Reviews* **3** (1978) pp. 30—45.
Woolgar, S.: 'On the Alleged Distinction Between Discourse and *Praxis*', *Social Studies of Science* **16** (1986) pp. 309—17.

Worcester Polytechnic Institute   STEPHEN J. WEININGER

does not acknowledge how much his own lucid aggression relies upon 'rhetoric': here, the exerting of pressure on common metaphors without drawing them to our attention. He thus reinforces assumptions and trains expectations. Those of us gathered as readers and writers of this volume are unlikely to see 'science' and 'literature' as merely competing modes, one of which must be expelled from the living space of the other. Indeed, recent new collections of essays emphasize plurality in their titles, as in *Languages of Nature*, edited by L. Jordanova, or inter-assimilation, as in *One Culture*, edited by G. Levine.

We may be far from the position of the alchemical writer Michael Maierus who saw his work *Atalanta Fugiens* (1617) as reaching its full scientific effectiveness only when it was read silently and aloud, its emblems pored over and copied, its instructions practically carried out and related to everyday life, and sections of it sung polyphonically. Such *Gesamtkunstwerk* is no longer part of the scientific enterprise nor even of the literary. Nevertheless, the imbrication of science and literature has not been obliterated.

The questions which weigh upon us are, far more than such totalizing, the *translatability* of concepts from one discourse to another, the trajectories along which ideas and metaphors move, and the degree of intercalation of scientific discovery and literary achievement. What are the *time zones* of reception and interaction? How do the significations of words shift, in time, across professional groups, across social classes, under what Virginia Woolf calls the "immense forces society brings to play upon each of us", (a phrase which effortlessly draws on a 'scientific' metaphor now become part of our repertoire of clichés)? Are such shifts stable, cumulative and irreversible, or are they constantly re-grouping? What are we to make of the urgent erasing of metaphorical congruities between diverse intellectual fields in what looks like an effort to reinforce their apparent autonomy?

Vigorous interchange of ideas and concerns between scientific and literary writers should not lead us to expect thoroughgoing and sustained congruity. Our analysis will founder if we seek, or expect, either a systematic representation of scientific ideas within works of literature or a sufficient 'source' of scientific theory within previous literary work. We are far more likely to find fugitive insight or generalized acceptance. For one thing, ideas do not remain constant when they change generic context. Yet in science and in literature the unspelled anxieties of the time may propel discovery (Beer, 1989a).

GILLIAN BEER

DISCOURSES OF THE ISLAND

1. THE QUESTION OF TRANSLATION

Since this essay was the outcome of an invitation from the Society for Literature and Science, let me begin my argument by dwelling on that copula, 'and'. Literature *and* science: science *and* literature: what is the force of the connective? It polarises the two domains; it yokes them together in a privileged pair, separated from other cultural expressions. It also sorts them hierarchically according to which is mentioned first, so that they become prime term and concessive term: *science* and literature; *literature* and science. One is given the originating role, the other that of dependent, providing 'context' or 'background'. Power struggles are masked by the deliberately evenhanded and non-directive 'and'.

Many writers have figured the relationship between the two activities as a territorial struggle. So, Peter Medawar in 'Science and Literature', his provocative Romanes lecture delivered twenty years ago quotes and inverts Lowes Dickinson's lament: "When science arrives, it expels literature". Medawar ripostes: "The case I shall find evidence for is that when literature arrives, it expels science". He accepts that "there are large territories of human belief" and learning "upon which both science and literature have very important things to say", but he then proceeds along the grain of the 'struggle for survival' metaphor that he has suggested in the words "expel" and "territories" and concludes with the selectionist "compete": "The way things are at present, it is simply no good pretending that science and literature represent complementary and mutually sustaining endeavours to reach a common goal. On the contrary, where they might be expected to co-operate, they compete" (p. 43).

One of his targets in that address was 'obscurity', as he saw it admired in then new French philosophical structuralism. His argument was the long-standing one that scientific language must be "as polite and as fast as marble" (a description he quotes from Joseph Glanvill in *Plus Ultra* [1668]). Medawar associates rhetoric with obscurity, and

had a similar importance in both. My instance, the island, has the advantage of giving prominence to Rudwick's further observation: "the concrete natural world does have an identifiable input, constraining though not determining the eventual outcome of the research" (p. xxii).

Ten years ago in 'Rhizome', Deleuze and Guattari articulated the problem of communality of understanding thus: "there is no language in itself, nor any universality of language, but a concourse of dialects, patois, slangs, special languages. There exists no ideal 'competent' speaker-hearer of language any more than there exists a homogenous linguistic community". They continue: "There is no mother-tongue, but a seizure of power by a dominant tongue within a political multiplicity" (p. 53). The writing here, it will be noted, harbors the glamorous violence it uncovers. The takeover bid, the military coup, disguises its power as maternal and originary; Deleuze's assertion has the same minatory force. On the other hand Deleuze's image of the "concourse" of dialects simplifies matters, for dialects do not only assemble but overlap, sometimes amicably, sometimes with cross-set significations which set up tensions in their reception. Discursive practices are not autonomous zones, nor do we operate within any single one of them alone. We habitually shift linguistic registers without difficulty. The island we set off for on holiday is other than the one we may inhabit, other again than the island economy of the Galapagos, or the theoretical zone generalized in population studies, other than Huxley's *Island* (1962) or Bacon's *New Atlantis* (1627), nor is it the place where Ariadne was abandoned. Each island is held within different discursive expectations. We are practiced in such transformations. We move freely among them. One of the singular powers of the mind (less often praised than its capacity for proliferating significations) is its power at will of excluding significations. But writers may choose to jar that easy passage across registers or open up the closed-off alternative meanings. Swift brings out in *Gulliver's Travels* (1726) how each new island requires a reconceptualization of the self for his hero. The island-idea is itself then isolated in that floating signifier, the flying island of Laputa.

Scientific revolutions do not often lead on to previously unimagined ideas; rather they discover previous imaginations and give them a demonstrable form which allows unforeseen implications to emerge. The imagination, unsustained by evidence, tends to repeat itself. Propelled by the forces of freshly combined fields of knowledge, statistics and genetics, geophysics and geology, new sequences of

Equally, the stories privileged in a culture tend to be privileged also in its scientific work (Holton). If simplicity, or hierarchy, or synchrony are valued they will be readily discovered (Beer, 1983; 1988a; 1989a). As workers in the field of 'literature and science/science and literature' we need to resist one-to-one systematization.

To recognize fugitive appropriation, or slack application, or covert story, without seeking to convert them into stable explanation or irreversible succession, does not at all mean that the critic's observations are without stringency. Quite the contrary. Such a position demands considerable knowledge and restraint — though it lacks the satisfaction and dignity of the stable completed pattern. Instead of seeking a single origin, or perfect gridding, the observer must scan a constellation of associated and unsettled material. That material will need to be drawn from the domains of scientific practice and theory, from diverse literary texts and narratives, and from other cultural formations. How much then are we falsifying or extending our understanding by the creation of a privileged pair 'literature and science/science and literature'? Such are the questions I want to address by means of my example of discourses of the island.

When the literary theorist I. A. Richards set out to describe and to distinguish between scientific and poetic language he called his essay, at first, *Science and Poetry* (1926). In later editions the title runs, less readily but more suggestively, *Poetries and Sciences* (1970): The maneuver brings out a central difficulty: is science a unitary system? — is literature? Much modern literary theory would deny that literature is an autonomous domain of writing; thus Stanley Fish in *Is there a Text in This Class?* argues that what will, at any time, be recognized as literature is a function of a communal decision as to what will count as literature. Could we say the same of science?

While literary theory in its post-structuralist phase has been emphasizing textuality (reading process, variety of discourse, illimitability of interpretation), writers in the history of science, such as Martin Rudwick in *The Great Devonian Controversy: The Shaping of Scientific Knowledge among Gentlemanly Specialists* and R. M. Young, *Darwin's Metaphor*, have demonstrated that scientific knowledge is the product of people in social interaction and have foregrounded the cultural or political constituents of such knowledge formation. The position of the 'author' has been differently treated in history of science and in literary theory but the emphasis on the role of cultural interaction has recently

Ariel poems harbors a double allusion: to that deep island exploration of bereavement, of possession, of colonialism, of the power of magic and reason which is Shakespeare's last play, *The Tempest*; and to the longed-for rediscovery of the daughter lost at sea and washed to unsafe shores which is the theme of Shakespeare's *Pericles*. But the line also stands bare of these humanistic concerns, its unpunctuated lapping rhythm expressing the touch of sea on land. Or we may recall the synaesthetic freedom of Hart Crane's line in 'Voyages II': "Adagios of islands, O my Prodigal". Almost unavoidable to the twentieth-century person practiced in reading within and outside literature, is, oddly, a fragment of a Donne meditation (XVI): "No man is an Island, entire of itself; every man is a piece of the *Continent*, a part of the *main* ... any man's death diminishes me, because I am involved in Mankind. And therefore never send to know for whom the bell tolls. It tolls for thee" (p. 538). This is a strange example of late fame — now even hackneyed, the sentences do not appear in nineteenth-century dictionaries of quotation. It is an extreme example of reading old texts into new history. It is, as it were, Eliot's reinvention, furthered by Hemingway's title *For Whom The Bell Tolls* (1940). So, though written in 1623, it becomes a twentieth-century response to the bleakness of Victorian individualism, represented, for example, in Thackeray's sentence in *Pendennis* (1850): "You and I are but a pair of infinite isolations with some fellow islands a little more or less near to us" (pp. 177–8). There, the promise of companionship in the word "pair" is flung away into "infinite isolations" and fellowship is transformed into a plurality of worlds (cf. Crowe).

Discourse does not take place outside history. But no historical period consists only of its own present. Not only architecture and legal systems give evidence of this, but — with particular intensity — past writing when read within and read into the present. History is in this sense less linear than constellatory. Whereas we skein out literary *production* into controllable periods — the Romantics — the Victorian Age — Modernism — *reading periods* are quite otherwise organized, trawling a variety of pasts and varying from person to person, though circumscribed by what is available within the community (Beer, 1985). In much recent discussion of the 'canon' and its powers the changing composition of the canon has often been overlooked: the interlocked accepted works of the late eighteenth century were very different from those of the late nineteenth century for example, and different again

probability emerge, to be gathered under titles such as plate tectonics and biogeography. I. Bernard Cohen in *Revolution in Science* specifies the task of his work as being "concerned with exploring and elucidating the creative process by which a practitioner in one discipline uses the ideas (concepts, methods, theories, tools) of another discipline" and sees the doctrine of 'transformation of ideas' as a key ingredient in the revolutionary process (p. xiii). I would extend Cohen's formula of creative appropriation beyond the sciences to include the 'concepts, methods, theories, and tools' of literature. I suggest, further, that it is not only *transformation* but the *recursiveness* of ideas that produces the significant interplay between literature and science. The very ease of transition across discursive practices raises the question of whether there is a shared and stable set of properties within the concept, a set which is trans-discursive, even trans-historical? Elements within an idea may be isolated and then extended in a new field of study. Such extension of specified elements from a concept goes beyond metaphor, without ousting irreversibly the previous set of interlocked ideas from which it derived. I shall demonstrate this later in relation to the idea of the island.

The idea of the island allows us at once the satisfactions of water and of earth, of deep flux and steadfast fruitfulness. At the same time it expresses the dreads of water and of earth, twin desolations, in which the self drifts or is confined, in which loneliness or loss predominate. The island has been persistently the experimental site, the enclosed zone for political utopias or dystopias. But the stable antinomies of earth and water have been disturbed in a number of ways in this century in our understanding of the island. How far has this changed the discursive and historical functions of the term? Do concepts change *irreversibly*, ('transform', 'revolutionize') under the pressure of new discoveries or ideas? Does not that assumption of irreversible change itself depend upon an unacknowledged evolutionary patterning? Is it not a self-conforming, because etymologically based, argument? A term may inhabit very different intellectual-historical periods simultaneously within a specific reader. (I. A. Richards simplified this problem by suggesting that large areas of our intellect resist change absolutely even while others move up to date.)

What flows into our minds from literature in the twentieth century from the single word 'island'? Perhaps, "what seas what shores what grey rocks and what islands". The first line of Eliot's *Marina* from the

ice on a river that is breaking up in the spring thaw ... Each continent does not constitute one plate, but rather each is incorporated with the surrounding ocean floor into a plate that is larger than the continent, just as a raft of logs may be frozen into a sheet of ice (p. v).

So the emphasis in plate tectonics is on fracture, drift, the lateral slide of plates against or alongside each other, the crumpling and overriding of one plate by another, the spreading and building up of ocean floors. The process described closely accords with another dominant theory of our own time: Derrida's concept of 'ungrounding' in epistemology (1976). The prolonged, casual mobility of the seemingly stable, the fragmenting of the originating universal-continent, the insistence on heterogeneity of effect, seem eerily close to the assumptions and representations of postmodernism both in fiction and philosophy. Yet it would be forcing the evidence, I believe, to propose a *causal* relationship in either direction between literary and scientific theory. Neither, though, is a *casual* relationship entirely convincing. Late-nineteenth-century geologists, in the midst of an engineering revolution which made extraordinary bridge building possible, favored the idea of sunken land bridges as their explanation of the connections between distant continental land masses. Current workers, sharing postmodernist terms, favor disgrounded plates, but have not abandoned entirely the inquiry into origins. Scientists, as well as the rest of us, think afresh with the metaphoric tools available within the current culture. Sometimes those tools open locked doors, sometimes they send us along a false trajectory.

To take an example of the difficulty of establishing a fit between discourses within a historical period: when I began this project I expected that psychoanalytic theory and dream repertoire would yield discussion of the hermeneutics of the island. Much of my reading indeed yielded for analysis hitherto undiscussed iconic representations of gender, and of gender formation, within the image of the island. But I gradually realized that Freud's topography is so preoccupied with border lands, frontiers, energy fields, and vertical sequences, that the island is of little conceptual use to him. Perhaps he entirely disbelieves in such separation in his spatial explanatory system. Jung's symbolic system privileges the oceanic and the submarine. The island is a defensive zone, a last outcropping of resistant consciousness, as he puts it, "a precarious island idyll": "The doctor knows these well defended zones ... ; they are reminiscent of island fortresses from which the

from our own institutionalized grouping of texts. This excursus explains why we may misconstrue the relations of literary and scientific writing within a period if we analyze *solely* the current social changes of that time. Past writing is always also part of the present day, newly inflected by being read and lived within fresh circumstances.

Let me take a striking example. Plato's island continent Atlantis is not so remote from Wegener's original continental land-mass which he names Pangaea in *The Origin of Continents and Oceans*. He names the sea surrounding it Panthalassa. Atlantis and Pangaea share conceptual characteristics, though they are remote in period of composition and in the social and scientific assumptions within which each writer worked. This is not at all to imply that Atlantis represents some real lost island-continent, or that Wegener's initiating idea, which led to the profound revolution in earth sciences known as the theory of continental drift or plate tectonics, is summarized only in his chosen mythical Greek names. Those names, however, make their own mythic claims. Pangaea asserts universality and totality: it means 'all earth', and it suggests all-embracing theory.[1] The collapse of good government in Plato's myth led on to the sinking of Atlantis; in Wegener's account the energies and the drift of earth broke apart the first earth mother or massive land form. The cosmogony in both cases links initiating great size and completeness to later loss and break-up.

N. L. Wegener's work on *The Origin of Continents and Oceans* (1915) did not succeed in convincing his contemporaries, but in the last fifteen years it has been reassessed by oceanographers, geophysicists and geologists. As a result of their joint studies they have come to accept what appeared in the past as an oversimple idea that, like an enormous jigsaw, the present continental forms of the globe originally fitted together in one vast continental island, surrounded by shallow waters, which Wegener called Pangaea and Panthalassa. One of the two broken continents, Laurasia and Gondwanaland, is given a Hindi name from a rock formation in India. Wegener was persuaded, not only by the physical fit of now widely distant coastlines but also by the evolutionary fit of the fossil record, that continents had broken up out of Pangaea and were now drifting laterally on plates. Tuzo Wilson explains it by means of a vivid metaphor:

No longer is the earth thought of as one rigid body with fixed continents and permanent ocean basins, rather scientists now consider the earth to be broken into six large plates and several smaller ones, which very slowly move and jostle one another like blocks of

and diversity of theoretical work which has made of it not only site but topic.

In literature, the island has been for far longer the space for exploration, self-inquiry, and satire of the writer's own culture. As landfall for adventure, islands are essential equally to Homer's *Odyssey* and to Book V of Rabelais's *Pantagruel*. They admonish present social organizations in More's *Utopia* (1516), Bacon's *New Atlantis* (1627), and later in Swift's *Gulliver's Travels* (1726). Island stories serve to justify or repudiate colonialism. In *The Tempest*, survival and inheritance are equally themes. Islands may provide beneficent retreats, figuring natural innocence, as in Saint-Pierre's *Paul et Virginie* (1788); or they may, as in the work of William Golding, figure first the breakdown of ordering values, in *Lord of the Flies* (1954), and then, more sardonically, the solipsism of human greed. Pincher Martin, shipwrecked on a rocky island, discovers at last that the island is his own tooth (1958).

It is no accident that first-person narrative is so frequent in literature of the island: the inner and outer are dangerously akin in solitude. The psychic space of the island is occupied by its describer. *Robinson Crusoe* (1719) is the generative narrative here. In Defoe, the island is an experimental site on which we watch the hero reformulate just such a bourgeois culture as he has sought to flee. The setting becomes not only the place but the condition of the experiment. Defoe's island is 'desert' — not unfruitful, but solitary. He must build up alone all the communal products of his civilization: bread, pots, fathers, and servants. Robinson Crusoe, for all his earlier rebellion, has no counter-imagination. Defoe de-natures language and the taken-for-granted by making us survey the complex production of humdrum life.[2]

The study of islands has been of the greatest importance in the development of evolutionary theory, from Darwin's theory-precipitating visit to the Galapagos Islands and his work on *The Structure and Distribution of Coral Reefs* (1839), his novel distinction between continental and oceanic islands, through to the articulation of the idea of 'natural selection' in *The Origin of Species* (1859). Not only Darwin but other nineteenth-century scientists in his circle saw the importance of the island for the pinpointing of evolutionary enclosure and change, and as providing historical evidence, and evidence for population studies. Hooker's lecture to the British Association for the Advancement of Science on the origins of island flora (1866) and Wallace's *Island Life* (1880) are important examples of these concerns.

neurotic tries to ward off the octopus" (*The Practice of Psycho-Therapy*, vol. 16, p. 181). In *Dream Analysis* he identifies "The Island of the Blessed", which he also calls "Atlantis", with the Unconscious (vol. 12, p. 149). Both Jung and Freud could draw on the strengthened awareness in late-nineteenth-century geology that islands are not autonomous but represent the *emergence* of submarine forces. At the time that Freud and Jung were writing, Wegener was first advancing the theory of continental drift. But they did not use Wegener, though he was there alongside them. Instead the nineteenth-century geological theory of sunken land bridges was sufficient to establish the degree to which land masses are mainly submarine, and continental islands are the outcropping of what "lies concealed beneath the [sea's] reflecting surface" (*Psychology and Alchemy*, vol. 12, p. 48). Or as Wegener put it: "Islands are usually larger fragments of continents, their substructures extending to about 50 km beneath the ocean floor" (p. 39). Wegener's time was yet to come and required new technology as well as new ideology for its explanatory power to be realized. As metaphor, indeed, the theory seems as yet not to have moved far beyond its own explanatory domain, save in the example of Charles Olson's *Maximus* poems.

2. INDIVIDUALISM AND THE ISLAND

Islands have become important in a range of scientific inquiries over the past hundred and fifty years or so. These outcropping land masses are assimilable to many intellectual, political, emotional, imaginative needs: brought under control by the mind through geography, geophysics, geology; used as experimental sites for political hypotheses; reached at last as refuge from the sea; evaded by the landlocked prisoner; affording an extremist miniaturized version of the earth's ecological ills; providing consonance with the human body and the human ego; an amphitheater of natural forces; and — according to Mandelbrot in *The Fractal Geometry of Nature* — an infinite coastline, crenellated beyond computation: "As increasingly small rock piles become listed as islands, the overall list lengthens, and the total number of islands is practically infinite. Since earth's relief is finely 'corrugated', there is no doubt that, just like a coastline's length, an island's total area is geographically infinite" (p. 116). We need think only of the island's role in evolutionary theory, population studies, ecology, anthropology, fractal geometry, and plate tectonics, to become aware of the richness

and tiny possessions took on the cultural characteristics of the nation which ruled them; of Ascension Island he writes:

> The whole island may be compared to a huge ship kept in first-rate order. M. Lesson has remarked with justice, that the English nation alone would have thought of making the island Ascension a productive spot; any other people would have held it as a mere fortress in the ocean (1839, p. 528).

The study of islands by Darwin, Hooker, Wallace and their contemporaries was undertaken at a time of extreme self-consciousness concerning England's own island status, then increasing into a far-flung empire along her many seaways. H. G. Wells, indeed, later characterised the British Empire as a 'steam-ship empire' whose decline began with the coming of the airplane. In the poetry of the mid-nineteenth century the connection is drawn tight between the island and the individual. In evolutionary theory also there is a considerable emphasis on the value of the individual exemplar as the bearer of genetic change. This emphasis draws, without analysis, on the expectations of the culture out of which it is produced, but has not yet been shown to be any less correct as scientific observation on account of its incorporation of cultural conditions.

Whereas in Darwin, Hooker, and Wallace, the emphasis in the island concept is on resourcefulness, productivity, and strangeness, in the poetry of the time it is upon loneliness, tedium, and lost community. Matthew Arnold's is perhaps its most famous expression. In two poems of the 1850's, "Isolation" and "To Marguerite — continued" he broods on separation:

> YES! in the sea of life enisled
> With echoing straits between us thrown,
> Dotting the shoreless watery wild,
> We mortal millions live alone.
> The islands feel the enclasping flow,
> And then their endless bounds they know.
>
> But when the moon their hollows lights,
> And they are swept by balms of spring,
> And in their glens, on starry nights,
> The nightingales divinely sing;
> And lovely notes, from shore to shore,
> Across the sounds and channels pour —

Both the extreme antiquity of preserved life forms in an island habitat and the evolutionary diversification of island life showed Darwin's concept of 'natural selection' at its most benign. The inhabitants of islands discover and occupy an extraordinary range of ecological niches so long as their economy is not disturbed by the powerful new resources of intruders. Darwin's chosen term 'intruders' signals the identification he makes with the 'inhabitants'. He takes the example of a country undergoing climatic change and contrasts the ecological effects on areas with open borders and on islands:

If the country were open on its borders, new forms would certainly immigrate, and this also would seriously disturb the relations of some of the former inhabitants. Let it be remembered how powerful the influence of a single introduced tree or mammal has been shown to be. But in the case of an island, or of a country partly surrounded by barriers, into which new and better adapted forms could not freely enter, we should then have places in the economy of nature which would assuredly be better filled up, if some of the original inhabitants were in some manner modified; for, had the area been open to immigration, these same places would *have been seized on by intruders. In such case, every slight modification, which in the course of ages chanced to arise, would tend to be preserved; and natural selection would thus have free scope for the work of improvement* (emphasis added) (1968, p. 131).

The activity of *natural* selection merges here with the idea of the 'natural' as the unspoiled. 'Improvement' is set alongside 'free scope', suggesting proliferation of possibility instead of curtailment. The harboring of early life forms within the island ecosystem emphasizes the ab-original. The island can accommodate slow change, diversification, and 'improvement' while still sustaining the early history of life on earth. The island thus becomes an originary image, bearing the force of Darwin's particular use of the word 'origins' to signify process rather than stasis. Its continuity and its physical enclosure preserves a full history of life forms into the present day.

Darwin, on the voyage of the Beagle, had objected to the hefty poeticizing of islands in contemporary descriptions. He writes of St. Helena:

A modern traveller, in 12 lines, burdens the poor little island with the following titles — it is a grave, tomb, pyramid, cemetery, sepulchre, catacomb, sarcophagus, minaret, and mausoleum (1839, p. 521, fn.).

But he noted with some English satisfaction the way in which remote

upon the island, on which she has scarcely leg room, so fully does she occupy its desolate space. Far off, in the corner of the picture, a minute ship sails away. Large woman, small island, distanced man: the icon pays a grotesque tribute to the heroic stature of her suffering and her enclosure. (Saltonstall's translation was made specifically so that women could read the stories.) Monique Wittig's *Les Guérrillères* has recently turned the tables on these images of islanded women.

Among Victorian writers Tennyson showed a particular sensitivity to this implication of island discourse: the melancholy of those left behind, stranded, chaste, half-sick of shadows, like the Lady of Shallott and Mariana of the Moated Grange. For him the image was not peculiarly that of the condition of women, though their enforced idleness doubled the tedium of introspection. The perfection of the self, the lack of replenishment from without, was one of his chief fears (as in "The Palace of Art"). In "The Lotos Eaters" he entrancingly presented the allure of the life without purpose, perfectly sated by its own completeness: yet to be feared.

In Victorian literature it is possible to imagine ideal islands, as Tennyson himself does in the death of Arthur and his withdrawal to the Isle of Avalon — but they tend to be islands of suspended growth, sufficient but interminable. In the scientific writing of the same time, however, the emphasis is on the burgeoning variety of flora and fauna on islands, the preservation of primitive forms, the extremes of large and small on them, the opportunities they give for minute variations to flourish without encroachment, the abundance of ecological niches discovered by the inhabitants, the energy of self-discovery.

In literature in the second half of the nineteenth century the *isolation* implied in individualism is emphasized — as well as its monstrous self-engagement. One of the best comic versions of this newly godless self-satisfaction (in which the individual simply annexes the enhanced psychic space of the island as his own), is W. S. H. Mallock's *The New Paul and Virginia, or Positivism on an Island* (1878), which has fun with the 'missing link' and above all with the hubristic notion of self-sufficiency. Shipwrecked on an ideal island the agnostic preacher greets his surroundings with "Oh important All! oh, important Me!" (p. 37). A few years earlier Browning in *Caliban upon Setebos, or Natural Theology in the Island* analyzed the imagination in the toils of self-explanation. Caliban extrapolates God from himself and his condition — a condition which emphasizes the random in the evolutionary, the

> Oh! then a longing like despair
> Is to their farthest caverns sent;
> For surely once, they feel, we were
> Parts of a single continent!
> Now round us spreads the watery plain —
> Oh might our marges meet again!
>
> Who ordered, that their longing's fire
> Should be, as soon as kindled, cooled?
> Who renders vain their deep desire? —
> A God, a God their severance ruled!
> And bade betwixt their shores to be
> The unplumbed, salt, estranging sea. (pp. 124—5)

Arnold here looks back longingly to the time before England had become an island isolated from the larger land forms. Edward Forbes at the 1845 BAAS meeting had advanced his theory "that the Continents once stretched across the ocean so as to include the islands" (pp. 62—8). The "ancient connection of the out posts or isolated areas with the original centres" that Forbes proposed became a metaphor for a lost relationship between individual and community in Victorian literature.

The double nature of the island — both a fragmentary upheaving of land from below the surface and a complete and autonomous form — is part of its imaginative attraction and makes it possible to play many nature/culture variations within its zone, all of them much explored by Victorian writers. Some of these variations are of great antiquity. The savage, the woman, the island, the lost paradise: early in the nineteenth century Byron mockingly noted how, for the shipwrecked hero Don Juan, the maiden Haidee comes with the island, hardly to be distinguished from it; there, like the land, to receive, serve and succor the incoming man (pp. 255—9). The identification of woman and island has been curiously little commented on, though it is there already in the Atlantis myth, where Poseidon places his human lover upon the central island and surrounds it by a series of alternating circular land and sea fortifications. The identification emphasizes the enclosure of the woman — often her painful abandonment, so that the island becomes simply a container. In Saltonstall's seventeenth-century translation of Ovid's *Heroides* (1671), for example, Ariadne is pictured after she has been abandoned: the distraught imposing figure of the woman stands forlorn

Daughter, Jane Gardam writes the social and psychic history of the twentieth century through a woman to whom nothing happens and by whom everything is experienced. Gardam substitutes the yellow house in which her heroine dwells from youth to extreme old age for Crusoe's island. The house, at first isolated on the edge of salt-marshes looking out to sea, ends surrounded by chemical plants and nuclear installations. The vehicles whirl around it: it has become a *traffic-island*, as absolutely isolated as Crusoe's isle.

One change in the idea of the island in recent years has taken place in the special discourse of biogeography. This discourse has unyoked the pair: land and water. That pair has been a basic tension in the idea of the island. The *Oxford English Dictionary* makes it clear that the word itself includes the two elements. 'Isle' in its earliest forms derived from a word for water and meant, 'watery' or 'watered'. In Old English 'land' was added to it to make a compound: 'is-land': water-surrounded land. The idea of water is thus intrinsic to the word, as essential as that of earth.

The equal foregrounding of land and sea is crucial not only in understanding the uses of the concept in imperialism, but in the more hidden identification between island and body, island and individual. The tight fit of island to individual to island permits a gratification which may well rely not only on cultural but on pre-cultural sources. The unborn child first experiences itself as surrounded by wetness, held close within the womb — though it is not an island in the strict sense since it is attached to a life-line, an umbilical cord. It becomes insular at birth.[3]

In ecology and population studies, the strand of meaning: *enclosed habitat* has been emphasized to the degree that it has become the basis for a revised topography of 'island'. In such discussion an island can signify a quite different series of antinomies from the old pair of land and sea. For example, in his book on *Island Populations*, Mark Williamson sets the scene thus: "The islands studied have been not only islands in the sea, but also islands in inland waters, and island-like habitats such as mountain tops, lakes, and even patches of vegetation, and individual plants" (p. v). In ecological terms, he points out, the lake around an island is itself an island: that is, it is an area which provides sustenance for its inhabitants and beyond which they cannot range and survive. The constraints of living needs which enclose species within

fickle power of a being struggling to discover a language by whose means he may escape his own abasement. Caliban is both primitive and modern man; the island is both Prospero's realm and the England of the 1860's, replete with imperial self-justification and gnawed by the doubts induced by evolutionary writing.

> — What Prosper does?
> Aha, if He would tell me how! Not He!
> There is the sport: discover how or die!
> All need not die, for of the things o' the isle
> Some flee afar, some dive, some run up trees:
> Those at His mercy, — why, they please Him most
> When ... when ... well, never try the same way twice!
> Repeat what act has pleased, He may grow wroth (p. 842).

The poem hints its doubts, too, about the process of colonization and about that metaphor embedded in mid-nineteenth-century evolutionary theory in which European man is the adult, while all other races, and women, are still children. The hazards and opportunities of individualism are at the explanatory center of evolutionary theory and of Victorian literature.

3. ECOLOGY AND ETYMOLOGY OF THE ISLAND

The word 'island' has a peculiar force in English which emphasizes its connection with individualism. The sounded 'I' at the beginning of the word creates a habituating consonance between the ego and the island. The New Zealand writer, Janet Frame, calls her autobiography *To the Is/land*. By dividing the word in two she emphasizes *Is* or *I*: Dasein or Ego. These potential senses surface only intermittently, but are always implicit in English discourses, in a way that they are not in *Insel* or in *île*, for example.

The island — the pocket — the house — the circle — the individual — the literary canon — the theater — the book: with varying degrees of extension all these concepts overlap with that of the island, exaggerating one or another characteristic to form a new topography. What we call '*pockets of resistance*' the French call '*îlots de resistance*'. In *The Poetics of Space* (1974) Bachelard opens out the significant enclosures of the house, recalling for us (though he does not mention it) that in classical Rome blocks of flats were called 'insulae'. In her recent novel, *Crusoe's*

Animals of all descriptions roved about in savage tranquillity. Here an enormous rhinoceros stood devouring the thorny leaves of the cactus; there a herd of elephants marched forward slowly and ungracefully, as if wearied with the weight of their colossal bodies; further on a troop of monkeys jabbered and chattered as they swung themselves on the lofty trees, and tried to rival each other in agility. Charmed with the picturesque scene, my wife and I sat down to rest and talk of the goodness of God, who had created such glories for the enjoyment of man. Meanwhile the boys prowled around with their guns, and an occasional shot told us that their search after game was not fruitless (p. 346).

There, in a particularly unselfconscious and popular form, is the old view of human dominance over other creatures. Now, so far has the wheel turned that it is necessary for an ecologist like Alison Jolly, whose work is predominantly with lemurs in Madagascar, to insist in her study of that island that human beings and their needs must be recognized as an authentic element in any ecological balance. The privileging of the human, the separation of observer from material life surveyed, has now — in claim, at least, though we may think not in economic actuality — become disgraced.

Islands in our present world history are the scene of some of the fiercest struggles between ethnic or political groups: Sri Lanka, Northern Ireland, Fiji, Cyprus (though when we look across the world we cannot, unhappily, suggest that only islands have these problems). The closed borders of the island state emphasize difference, which in human terms tends always to be conceived as cultural difference whatever its economic bases. The values of competition and selection which have haunted the last hundred years and claimed ratification from one reading of Darwin are not simply derived from evolutionary theory. Nevertheless, they share the less subtle evolutionary assumptions on which so much of our social patterning is based.

4. CONFLICTS AND ANXIETIES

We cannot *derive* literary from scientific revolutions directly: postmodernism is collateral with plate tectonics, not dependent on it. Nor can we *depend* on literature to provide blueprints or even harbingers of future scientific developments. The relationship between these diverse human activities is deeper than that and more mobile. Both draw on, are to some degree controlled by, and in their turn help to form, the common anxieties of the time. Gerald Holton speaks of the common assumptions of a period as its 'themata': *anxiety* needs to be added to

their habitat produce 'islands' whether surrounded by water or not. The impossibility of survival elsewhere exaggerates the element of enclosure already lodged within the concept.

In *The Fragmented Forest* Larry D. Harris points out that the island may be part of forest management: "The term 'island' spans the whole gradient from a small, distant atoll surrounded by a hostile medium (the sea) to a wilderness area that is defined by nothing more than a legal description in legislative minutes" (p. 80). The characteristic of the island concept conserved in island biogeography is that of *difference* from the surrounding environment, as in Frederick Gehlbach's title for his study of the borderlands of California and Mexico: *Mountain Islands and Desert Seas*.

The *distinctiveness* of the island habitat makes of it both a fruitful ecosystem for a limited range of species and an enclosure which may speed extinction. The emphasis on distinctiveness and enclosure underlies the seminal work of MacArthur and Wilson, *The Theory of Island Biogeography*, whose original title *An Equilibrium Theory* suggests a balance between immigration and extinction. Extinction happens more rapidly within the limited and unreplenished populations of islands than it does where there are open borders. Immigration to remote islands will be chancy and fitful, dependent more on rafting and on light body weight than on any purposive colonization. The discourse of such island study is plangent with elegiac terms. The subheadings of one of Williamson's chapters, "Features of island life", form a lapidary Hardyesque list: "The biota of Tristan da Cunha; Impoverishment; Disharmony; Dispersal; Loss of dispersal ability; Reproductive change; size change; extinction; relicts; endemics" (p. 30). A graveyard litany! — though it proves that not all these headings have quite the melancholic signification we might ignorantly give them, entering this island of discourse from another ocean. Relicts, for example, are here not widows but early or primitive forms that have survived on islands.

There is a remarkable contrast between the lopsided and often impoverished ecology of the typical island form described in biogeography and the jubilantly fruitful islands of much earlier literature. One of the best known nineteenth-century island romances is *The Swiss Family Robinson* (1813). Here, mankind — for this is the father's story, told by him — is fully justified in possessing — even, as we would see it, in pillaging natural life:

Utopia in 1838, J. A. St. John remarks "Sir Thomas More, influenced by partiality for the condition of his native country, erects his commonwealth upon an island — a position more favourable to independence and freedom than any other, as Pericles clearly intimates in Thucydides" (p. xliii). Shakespeare's *Richard II* provided the common self-description which still haunts our culture:

>   this sceptred Isle
> This other Eden, demy paradise,
> This fortresse built by Nature for her selfe,
> Against infection, and the hand of warre.

The island of England is seen as supremely and reflexively natural, "built by nature for herself".

The identification of England with the island is already, and from the start, a fiction. The island is equalled with England in the discourse of assertion, though England by no means occupies the whole extent of the geographical island; Scotland and Wales are suppressed in this description and Ireland is corralled within that very different group 'the British Isles'. Ireland is the necessary other, and has been so for centuries, in the English description of England. It is "John Bull's Other Island", part of which is determinedly *not* John Bull's. When Virginia Woolf wanted to summon the past of English life in her final work, *Between the Acts* (1941), at the onset of a war which seemed to spell the end of that old civilization, she imagined a house, a village, a pageant, a small girl who pipes

> This is a pageant, all may see
> Drawn from our island history.
> England am I . . . (p. 94)

And promptly forgets the rest of her lines. Woolf's revisionary pageant history excludes the empire and the army and by doing so calls attention to the expansionism so often politically associated with the island. Such expansionism still reappears at times in English life in the mythologizing of other far-flung islands, such as the Falklands or Fiji — a point eloquently made by Jonathan Raban, whose work I read after preparing this paper. Now, however, we in England are politically part of Europe and about to be joined to it, by a channel tunnel, by a mixture of science and technology and politics. We shall no longer be absolutely an island. Instead a tunnel, an umbilical cord, will re-attach us to that continent whose pre-history old Mrs. Swithin in *Between the*

that account. The island is a concentrating site for such studies in our own period. *Extinction* is one such anxiety: a massive generalizing of the individual fear of death which can operate in population studies, in ecology, in medicine, in post-nuclear literature, film and science fiction. In such anxiety even Noah's Ark can serve anew as the extreme image of the island population, consisting entirely of immigrants. A floating island, handmade, homely like a house, exclusive — leaving behind unicorns and gossips — dense with quarreling and energy. The medieval York mystery plays favored this pseudo-island for their drama. Very recently, another Yorkshire poet, Tony Harrison, has turned to look again at the Ark in a poem in a series "Art and Extinction" which seizes the deep connection between the death of a language, individual death, oppression, and extinction of a species:

> t'Ark
>
> Silence and poetry have their own reserves.
> The numbered creatures flourish less and less.
> A language near extinction best preserves
> The deepest grammar of our nothingness.
>
> Not only dodo, oryx and great auk
> waddled on their tod to t'monster ark,
> but 'leg', 'night', 'origin' in crushed people's talk,
> tongues of fire last witnessed mouthing: *dark*!
>
> Now when the future couldn't be much darker,
> there being fewer epithets for sun,
> and Cornish and the Togoland *Restsprache*
> name both the animals and hunter's gun,
> celebrate before things go too far
> Papua's last reported manucode,
> the pygmy hippo of the Cote d'Ivoire,
> and Upper Guinea's oviparous toad —
> (or mourn in Latin their imminent death,
> then translate these poems into *cynghanedd*) (p. 178).

The ark here doubles as the ship of death, bearing away the conquered.

I come from an island, an island which has insisted on a special status — moral and political — which its form has been seen (at least by the islanders) as authenticating. In his preliminary discourse to More's

between totalizing systems and local knowledge which have been a topic of much recent theory.[4] It emphasizes both inhabiting and observing. The observer comes in upon a complete world secured within natural boundaries; the island can be observed fully only by inhabiting it. In anthropology, where there is a persistent trouble about how to sustain the necessary roles of observing and of participating, the island site has been of particular importance. Partly this has to do with professional delimiting: we know where the edges of an island are and that makes for a tidier project.

But the equivocality of the inside-outside relationship of island and description dramatizes discursive doubling. Self-description and description of the other are hard to pry apart. First person narrative reinforces such doubling; equally, the apparently objective propositional prose of the scientist takes it for granted. Both serve to conceal a dilemma: the observer must learn to integrate as an inhabitant and yet must control a discourse beyond the reach of inhabitants. For no community can be sufficiently described within the terms at its own disposal. Yet the role of the observer is less flattering to the incomer than that of the accepted inhabitant.

Let me conclude, therefore, with an example from *Tristes Tropiques* that shows these tensions at work within the writing of Lévi-Strauss.[5] In order to control them he introduces the idea of the island into an account of his journey through tropical jungle. He even entitles the whole chapter "Robinson Crusoe" and in it he reaches the paradisal landscape which becomes the central place of his exploration. In this place axes give way and are flouted. Time is done away with. The scene is one in which he discerns sympathy between all living creatures, an ecological heaven of multiformity "marked by a tender intimacy between plants, beasts, and men". The chapter opens thus:

We had been travelling upstream for four days; there were so many rapids that we had to unload, carry and reload up to five times in the same day. The river flowed between rock formations which split it up into several arms; the reefs in the middle had trapped whole trees floating downstream, together with earth and clumps of vegetation. The vegetation had taken root so rapidly on the improvised islands thus created that it remained unaffected by the chaotic state in which the latest flood had left it. Trees were growing in all directions and flowers bloomed across waterfalls; it was impossible to say whether the river served to irrigate this fantastic garden, or whether is was about to be choked by a proliferation of plants and creepers, which seemed free to develop not only vertically, but through all spatial dimensions, as a result of the abolition of the usual distinctions between earth and water (p. 432).

Acts voluptuously imagined: "'Once there was no sea', said Mrs. Swithin. 'No sea at all between us and the continent'" (p. 38). The mythic discourses of the island are shifting now in England, yet evasively, without clear-cut equivalences to the coming loss of autonomy.

To give this paper I traveled from an island to the continent of America. I came by airplane. And it is the technology of the airplane which has most changed the island concept in our century. The island is no longer a fortress, defended by the sea. The axes are changed. It is overflown, lying spread out beneath the plane's surveying eye. We look down on islands now more often than across to them. The airplane has dislimned the tight boundaries of the shoreline; the sweeping eye of the observer moves in a single motion across land and sea. The airplane has made remote islands accessible and has resulted in their commodification as leisure package. The modern hotel waits to receive the traveler, the hotel itself a safe island of expected comfort within the sought-for wildness. But the commodification itself depends upon the unchanged survival of a long-standing ideal of the island.

The destructiveness and the new beauty generated by the possibilities of flight is realized by Gertrude Stein in her book, *Picasso* (1938) as Stephen Kern has also noted. In it she comments on the formal reordering of the earth when seen from the airplane — a reordering which does away with centrality and very largely with borders. It is an ordering at the opposite extreme from that of the island, in which centrality is emphasized and the enclosure of land within surrounding shores is assumed. Stein writes of the 1914—18 war thus: "Really the composition off this war, 1914—18, was not the composition of all previous wars, the composition was not a composition in which there was one man in the center surrounded by a lot of other men but a composition that had neither a beginning nor an end, a composition of which one corner was as important as another corner, in fact the composition of cubism" (p. 11). The patchwork continuity of an earth seen in this style undermines the concept of nationhood which relies upon the cultural idea of the island — and undermines, too, the notion of the book as an island. Narrative is no longer held within the determining contours of land-space. Intertextuality thrives.

5. DOUBLING AND SINGLENESS: CONCEPTUAL USES OF THE ISLAND

The island is both total and local, seeming to reconcile the conflicts

Scientific and literary discourses do not fit, but overlap. They do not succeed each other temporally. If we attempt to translate a concept too directly from one discourse to another we shall fail to recognize the differing weighting of elements within it. If we attempt to grid them upon one another too exactly we shall be uselessly puzzled by a loss of signification. Or perhaps too readily satisfied with apparent identity between them. Concepts do not change along a progressive pathway. Recursiveness is equally significant for interpretation. Older ideas of the island survive alongside new knowledge and interpretation, interpenetrating them. Scientists do not expect the discourses, or the results, of diverse fields of study simply to merge. Nor should we do so in studying the relations of literature and science. It is by placing them athwart each other that new patterns are discovered.

NOTES

[1] B. Mandelbrot (p. 5) comments on the narrow range of discourses from which scientific nomenclature is ordinarily drawn: "The terms coined in this essay . . . [tap] underutilized Latin or Greek roots . . . and the rarely borrowed vocabularies of the shop, the home, and the farm. Homely names make the monster easier to tame! For example, I give technical meaning to *dust, curd,* and *whey*." Elsewhere (p. 125) he offers "Cross Lumped Curdling Monsters" and "Knotted Peano Monsters, Tamed". In *Fiasco*, S. Lem elaborates this idea, naming one phenomenon "Birnam Wood" and having a character comment: "Gorgon, Typhon — we're lucky the Greeks have so many monsters in their mythology for us to borrow" (p. 11). See also Beer (1985). Space has prevented me from entering here upon the topic of science fiction and the idea of plurality of worlds in relation to the island but this will form part of my work in progress in this field.

[2] Defoe's work is remarkable also for the way in which it has yielded a series of creative responses which bring to our notice one or another of Defoe's cultural assumptions and, equally, the characterizing anxieties of a later age. Most recently we have seen Michel Tournier's *Vendredi*, Jane Gardam's *Crusoe's Daughter*, and J. M. Coetzee's *Foe*, in which issues of gender or of colonialism are concentrated.

[3] Since completing this paper I have been pleased to find confirmation of the idea in Michèle le Doeuff.

[4] Kate Hayles has kindly allowed me to read a chapter from her forthcoming book on chaos theory which independently offers a substantial development of the argument here adumbrated concerning the tensions between local knowledge and global theory. I am grateful to her for the opportunity.

[5] For a different reading of this scene compare Jeffrey Mehlman, pp. 11—14.

REFERENCES

Arnold, M.: *The Poems of Matthew Arnold* (ed. by K. Allott), Longmans, London, 1965.

The river is multiform, the islands "improvised", "the usual distinctions between earth and water" are abolished. Polarities give way, classification crumbles, yet the images of island and of garden hold the scene together. The idea of the island allows redundancy to become rapture in Lévi-Strauss's heaving, profuse description. Life coheres within its zone. He needs to invoke autonomy and enclosure because the sustained emphasis of the chapter is on what he calls the earth's "virginity" within profusion. His description is highly sexualized. He must discover singleness within the polymorphic pleasures he observes:

> Where exactly does that virginity lie, behind the confusion of appearances which are all and yet nothing? I can pick out certain scenes and separate them from the rest; is it this tree, this flower? .... I reject the vast landscape, I circumscribe it and reduce it to this clayey beach and this blade of grass which is trodden daily by the most authentic savages but from which, however, Man Friday's footprint is missing (p. 437).

Lévi-Strauss needs particularity, above all. The single instance can then become the whole, not merely stand in for the whole. He needs it here, it would seem, for the sexual pleasure of himself and his reader. But he needs it argumentatively too. The eye of the observer can in honesty dwell only on the particular. That is another methodological reason for the constant reaffirmation of the island-idea across many intellectual fields. The intense focus of the observer's eye can be lodged so firmly on this miniaturized zone that she or he can claim simultaneously empathy and control. These twinned, often irreconcilable, ideals within the writing of the human sciences and of literature, can find a point of concentration in the single concept: island.

In Lévi-Strauss's ode Friday's footstep is declaredly absent. Thereby incursion is signaled and denied. The absence of the alien serves to conceal Lévi-Strauss's own presence in the text: it covers his tracks. In this *jouissance* he seeks to evanesce and yet to continue writing. The invoked metaphors of the island serve not only his rhetorical needs but those of workers in a number of fields. In *The Return to Cosmology*, Stephen Toulmin comments on the vanishing of the observer in recent scientific thinking. The observer has no privileged space, but is implicated, folded in. In such conditions island boundaries have a further significance, representing the discursive doubling of observing and inhabiting. They remain desired, and inscribed, despite the concurrent dislimning of its limits in wave theory, in plate tectonics, in fractal geometry, in the airplane's downward-gazing flight.

Islands, considered with regard to Geological Changes', *Transactions of the British Association for the Advancement of Science*, Cambridge University Press, Cambridge, 1846.
Forsyth, P.: *Atlantis The Making of a Myth*, McGill-Queens, Montreal and London, 1980.
Frame J.: *To the Is-land*, The Women's Press Limited, London, 1983.
Gardam, J.: *Crusoe's Daughter*, Hamish Hamilton, London, 1985.
Gehlbach, F.: *Mountain Islands and Desert Seas: A Natural History of the US-Mexican Borderlands*, Texas A & M Univ. Press, College Station, Tex., 1981.
Golding, W.: *Lord of the Flies*, Faber and Faber, London, 1954.
Golding, W.: *Pincher Martin*, Faber and Faber, London, 1956.
Harris, L.: *The Fragmented Forest: Island Biogeography and the Preservation of Biotic Diversity*, Chicago Univ. Press, Chicago, 1984.
Harrison, T.: *Selected Poems*, Penguin, Harmondsworth, 1984.
Holton, G.: *Thematic Origins of Scientific Thought: Kepler to Einstein*, Harvard Univ. Press, Cambridge, Mass, 1974.
Hooker, J.: 'Insular Floras', *Transactions of the British Association for the Advancement of Science*, Nottingham, 1866.
Huxley, A.: *Island*, Chatto and Windus, London, 1962.
Jolly, A.: *A World Like Our Own: Man and Nature in Madagascar*, Yale Univ. Press, New Haven and London, 1980.
Jordanova, L., ed.: *Languages of Nature*, Free Association Press, London, 1986.
Jung, C.: *Collected Works*, 20 vols., Rouledge and Kegan Paul, London, 1979.
Lem, S.: *Fiasco*, Andre Deutsch, London, 1987.
Lévi-Strauss, C.: *Tristes Tropiques* (trans. by J. and D. Weightman) Jonathan Cape, London, 1973.
Levine, G., ed.: *One Culture*, Univ. of Wisconsin Press, Madison, 1988.
MacArthur, R. H. and Wilson, E. O.: *The Theory of Island Biogeography*, Princeton Univ. Press, Princeton, 1967.
Mallock, W.: *The New Paul and Virginia, or Positivism on the Island*, Chatto and Windus, London, 1878.
Mandelbrot, B.: *The Fractal Geometry of Nature*, W. H. Freeman and Co., San Francisco, 1982.
The Dictionary of Imaginary Places (ed. by A. Manguel and G. Guadalupi), Granada, London, 1980.
Marvin, U.: *Continental Drift: the Evolution of a Concept*, Smithsonian Institution Press, Washington, 1973.
Medawar, P.: *Pluto's Republic*, Oxford Univ. Press, Oxford, 1982.
Mehlman, J.: 'The "Floating Signifier": From Lévi-Strauss to Lacan', *Yale French Studies* **48** (1972) pp. 10—37.
More, T.: *Utopia: or The Happy Republic. A philosophical Romance. To which is added, The New Atlantis, by Lord Bacon. With a preliminary discourse by J. A. St. John, Esq.*, Joseph Rickersby, London, 1838 (First edn. 1516).
Pearson, B.: *Rifled Sanctuaries: Some Views of the Pacific Islands in Western Literature to 1900*, Auckland Univ. Press, Auckland, New Zealand, 1984.
Raban, J.: *Coasting*, Pan Books, London, 1987.
Richards, I.: *Science and Poetry*, Kegan Paul & Co, London, 1926.

Bachelard, G.: *La poetique de l'espace*, 8th edn., Presses universitaires de France, Paris, 1974.
Ballantyne, R.: *The Coral Island: a tale of the Pacific Ocean*, T. Nelson and Sons, London, 1858.
Beer, G.: *Darwin's Plots: Evolutionary Narrative in Darwin, George Eliot and Nineteenth-Century Fiction*, Routledge and Kegan Paul, London, 1983.
Beer, G.: 'Darwin's Reading and the Fictions of Development', in *The Darwinian Heritage* (ed. by D. Kohn), Princeton Univ. Press, Princeton, 1985.
Beer, G.: 'Designing and Describing: A Problem in the Language of Discovery', in *One Culture: Essays on Literature and Science* (ed. by G. Levine), Univ. of Wisconsin Press, Madison, 1988(a).
Beer, G.: 'The Death of the Sun: solar physics and solar myth', in *The Sun is God: Essays on Nineteenth-Century Mythography* (ed. by B. Bullen), Oxford Univ. Press, Oxford, 1989(a).
Beer, G.: 'The Island and the Aeroplane: The Case of Virginia Woolf', in *Nation and Narration* (ed. by H. Bhabha), Methuen, London, 1989(b).
Browning, R.: *Poetical Works 1833—1864* (ed. by I. Jack), Oxford Univ. Press, Oxford, 1970.
Byron, G.: *Don Juan and other satirical poems*, The Odyssey Press, New York, 1935.
Coetzee, J.: *Foe*, Secker and Warburg, London, 1986.
Cohen, I.: *Revolution in Science*, Harvard Univ. Press, Cambridge, Mass., 1985.
Compère, D.: *Approche de l'ile chez Jules Vernes*, Minard, Paris, 1977.
Crane, H.: *Collected Poems* (ed. by B. Weber) Bliodaxe Books, Newcastle Upon Tyne, 1984.
Crowe, M.: *The Extraterrestrial Life Debate 1750—1900: The Idea of A Plurality of Worlds from Kant to Lowell*, Cambridge Univ. Press, Cambridge, 1986.
Darwin, C.: *The Geology of the Voyage of the Beagle*, Smith, Elder, and Co, London, 1842—6.
Darwin, C.: *Journal of Researches into the Geology and Natural History of the Various Countries Visited by M. M. S. Beagle, under the command of Captain Fitzroy, R. N. from 1832 to 1836*, Henry Colburn, London, 1839. Edition cited: T. Nelson and Sons, London, n.d.
Darwin, C.: *On the Origin of Species by Means of Natural Selection*, John Murray, London, 1859. Edition cited: *The Origin of Species* (ed. J. Burrow), Penguin Books, Harmondsworth, 1968.
Defoe, D.: *The Life and Strange Surprising Adventures of Robinson Crusoe of York, Mariner*, W. Taylor, London, 1719.
Deleuze, G. and Guattari, F.: 'Rhizome', *Ideology and Consciousness* **8** (1981) pp. 49—71.
Derrida, J.: *Of Grammatology* (trans. by G. Spivak), Johns Hopkins Press, Baltimore, 1976. See esp. 'The Supplement of (at) the Origin', pp. 313—16.
le Doeuff, M.: *Recherches sur l'imaginaire philosophique*, Payot, Paris, 1980.
Donne, J.: *Complete Poetry and Selected Prose* (ed. by J. Hayward), Nonesuch Press, Bloomsbury, London, 1929.
Eliot, T.: *The Complete Poems and Plays of T. S. Eliot*, Faber and Faber, London, 1969.
Fish S.: *Is There a Text in this Class?: The Authority of Interpretive Communities*, Harvard Univ. Press, Cambridge, Mass., 1980.
Forbes, E.: 'On the Distribution of Endemic Plants, more especially those of the British

Rudwick, M.: *The Great Devonian Controversy: The Shaping of Scientific Knowledge among Gentlemanly Specialists*, The Univ. of Chicago Press, Chicago and London, 1985.
Saint-Pierre, J.: *Paul et Virginie*, Lausanne, 1788. Trans. as *Paul and Mary. An Indian Story*, J. Dodsley, London, 1789.
The Dialogues of Plato (trans. by B. Jowett), Clarendon Press, Oxford, 1868—71.
Saltonstall, W.: *Ovid's Heroical Epistles*, William Whitwood, London, 1671.
Shakespeare, W.: *Richard II*, Cambridge University Press, Cambridge 1939.
Swift, J.: *Gulliver's Travels* (ed. by H. Davis), Basil Blackwell, Oxford, 1941.
Tennyson, A. Lord: *The Poems of Tennyson* (ed. by C. Ricks), Longmans, London, 1969.
Thackeray, W.: *Pendennis* (ed. by D. Hawes), Penguin Books, Harmondsworth, 1972.
Toulmin, S.: *The Return to Cosmology: Post Modern Science and the Theology of Nature*, Univ. of Californian Press, Berkeley and London, 1982.
Tournier, M.: *Vendredi, ou, les limbes du pacifique*, Paris, 1978.
Verne, J.: L'Ile mysterieuse, Paris, 1874—5.
Wallace, A.: *Island Life: or, the Phenomena and Causes of Insular Faunas and Floras*, Macmillan, London, 1880.
Wegener, N.: *The Origin of Continents and Oceans*, rpt. Methuen, London, 1966. (First pub. 1915).
Williamson, M.: *Island Populations*, Oxford Univ. Press, Oxford, 1981.
Wilson, A.: *Continents Adrift and Continents Aground*, W. H. Freeman and Co., San Francisco, 1976.
Wittig, M.: *Les Guérrilères*, Les Editions de Minuit, Paris, 1969.
Woolf, V.: *Between the Acts*, Hogarth Press, London, 1941.
Wyss, J. D. and J. R.: *The Swiss Family Robinson*, T. Nelson, London, 1852. (First pub. 1811).
Young, R.: *Darwin's Metaphor: Nature's Place in Victorian Culture*, Cambridge Univ. Press, Cambridge, 1985.

Girton College, Cambridge

mean privilege, in the sense of privileging this transitional era over others: the nineteenth and twentieth centuries are, after all, equally crucial for other discourses. Nor do I mean privilege in the sense of suppressing their contrary discourses (i.e. *counter*-nerve, *counter*-melancholy, all those supernatural discourses that discourage secular taxonomy and empirical observation) or privilege as achievement, in the sense that Cartesian dualism or Lockean associationism can be said to have altered the fundamental paradigm of mind and body the Enlightenment inherited from the Renaissance. Yet during this early period the four discourses of mind and body broke away from the older, supernatural forms and embedded themselves in new discursive forms and genetic structures, some of which will soon be called 'scientific' literary forms by the Royal Society; and they endured until the end of the Enlightenment, when another break — Romanticism (or call it Lamarckism or proto-Darwinism) — will again change the way the ampersand in mind *and* body is construed.

The reciprocity of mind and body is also intriguing as an analogue of the problematic in the ampersand of Literature and Science, even though there is no contemporary subject, no recognized field theory, of mind and body. One has to create a subject for it oneself: a narrative, a rhetoric, a story line, even to grasp what role the footnotes on mind and body can contribute to a field for it, just as Aldous Huxley, a much underestimated pioneer for Literature and Science, had to invent a narrative for a book replete with the discourse of mind and body, which he eventually subtitled "a study in the psychology of power politics and mystical religion in the France of Cardinal Richelieu", and whose main title is *The Devils of Loudun* (1952).

But what *is* privileged here are the competing discourses of mind and body, their overlaps and reciprocities, their mutual rememberings and forgettings, on the rationale that we might learn something by analogy from the ampersand in their conjunction, and — furthermore — because cultural history has never had a subjective and social, as distinct from a logical and internalist, history of their reciprocity (see Porter). So this approach through overlapping discourses entails Literature and Science as an antidote, at least, to the impoverishing incompleteness of existing historical explanation. When we consider the social antagonisms that mind and body have served, is it any wonder that we should want to inquire after their ideological relation? The profiles of mind and body have been widely disparate in the popular imagination,

G. S. ROUSSEAU

DISCOURSES OF THE NERVE

I call this two-part talk "Discourses of the Nerve", having developed it as part of a larger project on the discourses of mind and body on which I have been engaged for the last few years (Rousseau, 1975); but in different settings and circumstances than this one it could as well have been called "The *Product* of Literature and Science", for — as you will see in Part 2 — I am concerned as much with the *product* of our activity, with the finished object, the eventual discourse, the narrative produced at the end of the process we are beginning to call 'Literature and Science', as with the discourses themselves.

1. THE DISCOURSES OF MIND AND BODY

Between the late Renaissance and the start of the nineteenth century, the troubled theoretical relations of mind and body were encoded in dozens of different discourses. These were layered, or tiered, in the way Foucault and other contemporary discourse theorists have demonstrated (Foucault, 1973). Four of these discourses dominated the models of the early European naturalists, who were not yet scientists in our modern sense. I call these the discourses of melancholy, hysteria, the nerve, and sensibility, despite the vexed and thorny problems of definition and taxonomy, and the symptoms that were then diagnosed under the perplexing umbrella of hysteria (see Appendix A). Ultimately, all were discursive attempts to medicalize the imagination, then believed to be a crucial activity for the analysis of cognition and consciousness as well as mental pathology (Rousseau, 1975). Despite their differences, their overlaps and reciprocities, their rhetorics and their grammars, the four discourses had this medicalization in common; but no sooner were the discourses analyzed than the endeavor's implicit and inexorable metaphysic of mind and body became apparent. Before the early seventeenth century, mind and body remain demonically encoded; afterwards, they gradually reside in the domain of the naturalistic. Given the relatively swift diachronic transformation this is a crucial epoch for relations of mind and body. But by crucial I do not

philosophers demonstrate that the dualism of mind and body no longer exists today in any rigorous sense, on an important *non-philosophical* level the split is not a myth but an institutional reality. Until we can account for why this divide was produced and continues to be maintained in many quarters (just like the divide of Literature and Science); until we trace out what interests the proponents of mind versus those of body have had; we are unable to abolish its existence by fiat.

As a topos mind and body extended everywhere into Renaissance discourse: in sermons, in poems, in prose romances; in didactic treatises, in moral memoirs, in theological tracts; and — of course — in the anatomical and physiological discourses that figure in Appendix A.[3] The literary component of mind and body was also extensive. As the Renaissance waned, the four discourses gradually assumed the non-ecclesiastical (or secular) charge of the reciprocities of mind and body, just as colleges in our time have proved to be the landlords of arts *and* sciences (that problematic yet influential ampersand again) when viewed conjunctively.

1.1. *The Discourse of Melancholy*

Diachronically, the discourse of melancholy was the first to develop. It was an old discourse by the time Burton built his giant edifice on its ruins. Under less able pens or less polymathic minds than those of Burton in holy orders — minds like those of Timothy Bright, Andre Du Laurens and Felix Platerus (all physicians), or the divines who warned about the signs of religious melancholy — melancholy's rhetorical techniques and genetic encodings differed, but its semiotic remained the same. Melancholy took the chemical humors, especially bile and bilious fluids in the body, as its main sign, enlisting madness and derangement as its signifier, stressing that the sign was material substance and signifier predictable human behaviour.[4] To the degree that its authors were primarily physicians interested in therapy and cure, it is not surprising that sign and signifier should have been privileged in this way, nor odd that melancholy constructed a symptomology based on the derangements of the four humors by the determinants of climate, geography, terrain, weather, topography, and national boundaries. The diagnosis of melancholy was in this sense an interpretation of signs predicated on symptoms. Yet what is extraordinary about the *discourse*

body having suffered miserably at the expense of mind.[1] But body had been a trope for centuries; what changed in the Enlightenment was an altogether new level of metaphoric energizing: just the opposite of what the old school of historians of the Royal Society (R. F. Jones, Morris Croll, even Marjorie Nicolson) would have us believe about science and the 'plain style', about the constitution of a new genetic literary form called 'the scientific paper', or 'the scientific essay': plain and straightforward, without embroidering language in any way, laying out its parts like a mathematical proof — Q.E.D. — and then 'embodying' this thing, this amorphous knowledge called 'science'. I do not privilege either mind or body here as a completeness axiom for philosophers or historians of science but as an alternative approach to an inquiry they alone have mapped diachronically,[2] and I do so in the attempt to constitute a category, a subject, a field for it. Today it is a cliche to maintain that no one has construed this dualism seriously for over a century; it was not so before 1800, when both were terrifically mythologized. The ways they continue to be mythologized afterwards, is an area of debate this paper hopes to open up.

Anyone with an ear close to contemporary neurophysiology or sociobiology could surmise that there is no dualism: only the monism of matter and motion — vulgarly, that there is brain and essentially nothing but brain, as the philosopher Thomas Nagel has admirably argued (Nagel). But is brain mind or body? More importantly, is intellection mind or body? If my doctor tells me I must have a heart or kidney transplant, I do not think I will lose my identity as George Rousseau. But if my neurologist informs me that I must have a brain transplant — assuming we have reached a more advanced stage of medicine — I wonder what of me will remain, this despite the knowledge that the mental functions permitting intellection are not *all* located in the cerebellum. Here the logic gets thorny, if for no other reason than that the issue of selfhood and identity is intimately tied to these matters, as even the philosophers and scientists of early modern Europe recognized. No sooner do we claim brain is body, than the resonances of *brain as mind* clamor for attention on grounds that we cannot be identities that are less than our own consciousness. Many academic groups (not merely philosophers) privilege mind over body, even if they do not subscribe to its dualism. Among cultural and social historians it is clear that body has had a bad press since the Renaissance (see Porter; Gallagher and Lacquer; Bottomley; Turner), so that even if

had not yet been fixed, its agenda was egalitarian *vis-à-vis* gender and sex, especially because the Renaissance perceives human imagination as androgynous. There is no sense of a hierarchy of genders other than that which the culture inscribes. Unlike other discourses then, melancholy conceals no hierarchies *vis-à-vis* gender, any more than it veils the author's ideology. This is why, later on, Swift could model his *Tale of a Tub* and Sterne *Tristram Shandy* on a neutral Burton whose only ideological agenda is the genial madness these authors cultivate through their mad hacks and disoriented narrators. In Bright and Burton, in Du Laurens and Platerus, males as well as females are mad, and female bile remains the same as male. Unlike the status of hysteria, the agenda here entails universal structures, a human rather than a genderized humoral chemistry, with Hottentots and Indians, for example, as prone to this humoral chemistry and subsequent diagnostic as the Queen of Portugal and Prince of Wales.

The ideology of the doctors also merits attention: paradoxically, they claim to be inside and outside the discourse at the same time, enmeshed in the signs they read by virtue of their professional skill, but nevertheless apart from the madness diagnosed. (I have not yet found a single female physician who wrote about melancholy and therefore I continue to invoke the male pronoun.) He legitimizes his writing by implied beneficence: that somehow the whole culture benefits from his 'scientific' expertise because almost everyone is depressed. This legitimation reveals the sequence of his power: to see in order to interpret; to interpret in order to diagnose; to diagnose in order to cure. No one (high or low, male or female) is marginalized. His power lies at the center because the whole world is confused, chaotic, mad; and the physician has discovered, he claims, the objective source of the chemistry of madness. In bile (whether black, white, yellow, or green), not in the mystery of the Godhead or the elusive human imagination, in bile lies the secret of the human Book of Nature. Governments rise and fall, human are born to die, but bile, it appears, remains a constant: it is an extraordinary mythology, claiming to be as 'scientific' and 'dispassionately objective' as anything the Ancients and the Renaissance ever produced, but in fact it was rhetorical, polemical, ideologically loaded, anything but dispassionate. The medicalization of imagination was a groping for the unknown through the tangible of the bile. No wonder that Freud and the psychoanalysts of the early twentieth century revolted against this agenda and generated another myth of depression to replace it.

of melancholy — as distinct from individual symptoms or individual case histories — is the way melancholy was encoded in discursive genetic forms and then privileged among competing discourses as the embodiment of the cultural myth of mind and body. 'Embodiment' is, of course, a problematic concept, presenting many types of metaphysical hurdle to the theorist of discourse; it is not a problematic we can resolve here. But set the diachronic dials to 1550 or 1600, or even 1650: no other discourse of mind and body has been so universally mythologized.[5] Whether as Hamlet the gloomy Dane, or in Taylor's "Lamentation" of 1618 (see Taylor), melancholy captures the European imagination everywhere. As all words and things then revealed a morality and correspondence, so too did the melancholy of the doctors. It was the given belief that the malady resides as much in the mind as in the body. This dual residence is uttered in the aura of terror and wonder, as if this awareness had been the great scientific discovery. Doctors insisted that melancholy was no one pathological state but a more general condition of imagination. There were thought to be so many types of melancholy that it can be said of melancholy that it becomes the signifier of something *other* than biliousness; something akin to, but as yet not identical with, social status. This juncture is where the internal contradictions of the discourse become patent, for few human conditions have generated such metastories and scientific critiques.

Historically, both a literary and a scientific discourse never developed; the bifurcation is a figment of our post-Kantian imagination.[6] Dr. Bright learned as much classical rhetoric at Cambridge as any philologist; he invokes the trope of *partitio* as often as Burton, Bacon, and the non-physicians.[7] Burton not only heightened this use of classical rhetoric, but encoded his own encyclopaedic tendencies into the vehicle of the 'anatomy'. His *Anatomy* becomes the repository of the whole world (allegedly because bile and madness are endemic), encoded in a literary vehicle (the anatomy) whose essence lies in disjunction and fragmentation. The more Burton searches for the causes of the signifier madness, the less he can find them. All he discovers as he ferrets out truth from words and things, is chaos and confusion. In the process he himself seems to become mad and take on the attributes of the very signifier (madness) whose cyphers he claims to be decoding.[8]

But if the discourse of melancholy was *neither* essentially literary nor scientific, neither natural nor supernatural, because these intersections

the way it marginalized the female body, but he penetrated right through its claims to objectivity. "This inferior [female] body which Sydenham tried to penetrate 'with the eyes of the mind' was not the objective body available to the dull gaze of a neutralized observation" (Foucault, 1970, pp. 125—6). In the discourse of hysteria, this female body was a moral and ideological imagining. Foucault's study of medical texts led him to this insight. Yet, unlike the more egalitarian melancholy that privileged bile within the humoral derangement while devaluing the rest of the anatomy, hysteria was *par excellence* the suppression of the *female* body in her most vulnerable genital organ. Hysteria, the antidote to melancholy, was the discourse that dominated gender and sex in early Modern Europe, its concealed program the *inferior* (as well as interior) female body no matter to what degree it made altered states of consciousness its explicit agenda. Its more urgent subjects were the allegedly soft spaces and useless flaccid interiors of women (so much for scientific objectivity in the face of female bodies which, if they are anything, are more durable than male!). Thomas Sydenham, the English physician and medical partner of John Locke, will strive to alter this status in the 1680s for reasons that have not yet been explored (see Dewhurst, pp. 36—46). Sydenham will develop a new 'hypochondriasis' as the male equivalent of female hysteria and defend the bifurcation of this condition into male and female versions on scientific and objective grounds, whereas in point of fact the gender revolution of the Restoration lies behind his privileging of the male body in this new way. In brief: to be hypochondriacal in the 1680s and 1690s entails a new means of setting males apart from other males.

After ca. 1700 hysteria will again be transformed as its sign and signifier depart from its representations in the fifteenth century *Malleus Maleficarum*; its discursive monopoly withdrawn, so to speak, as it marginalizes *upper-class* women in the name of spleen, vapours, 'hyp' (the vernacular term in the eighteenth century for medical hypochondria), nerves. But until then — until Willis and Sydenham — the discourse of hysteria serves to explain to women, however diverse its rhetorical strategies, why they are failures as a result of their anatomy; more specifically, as a result of their irrational cravings, erotic appetites, soft daily routines, idleness, unrelenting boredom, and especially their capacity for sorrow and suffering. And it reassures men, conversely, that they are immune from such frailty, having been anatomically

1.2. The Discourse of Hysteria

Renaissance and early European hysteria differed, although its tropes were stylistically no less neutral than those of melancholy, no less teleological. But hysteria developed mechanisms of concealment for its teleology — this despite its value-laden, ideological, and rhetorical agendas and despite its own self-conscious claims to objectivity. In some ways hysteria was the nighttime of melancholy. Its ideologies thrived on distinctions of gender, its male proponents having privileged male anatomy over the allegedly weak female genitalia from the earliest known pronouncements at the time of Hippocrates and Soranus of Ephesus. We should probably not overlook within this context that not a single male received the 'stigma' (the sacred stigmata) between the Middle Ages and the twentieth century, between Francis of Assisi and Padre Pio, though hundreds of women did.

Like melancholy, the discourse of hysteria aimed to show how endemic among females the signifier of hysterical behavior was, but here similarity ended, its agenda and program having a different purpose, as did its discursivities and diachronic appearances. What the Greeks meant by the 'wandering womb' or the 'suffocation of the mother' (*hysterike pnix*) bears little resemblance to the medical hysteria of the Renaissance despite the conventional histories of Ilza Veith and, before her, in France, the larger one by Georges Abricosoff, histories that will now no longer do.[9] Until the Enlightenment hysteria is said to be the female condition of the 'wandering uterus', significant for medical doctors when disordered or deranged. This mania is the 'furor uterinus', the swelling of the so-called rising womb, the 'furor' that Enlightenment physicians will exalt as the cause of the newly defined condition nymphomania (the mania again delimited to women: 'nymphs'); not actually wandering like a displaced organ, or floating like a self-contained island throughout the female body and within its rivers of blood, but rising — it was then said by the doctors — so high within its own position and thrust so extensively into the territories and domains of the other organs, into their fissures and crevices, their unoccupied spaces and open fields (this open field is the metaphor that continues to be used), that one might as well consider it a 'wandering uterus'. When inflamed and swollen it will be described as 'suffocating', stifled, choked; especially on the spatial grounds that there is no unoccupied vacancy left for its swellings to permeate.

Foucault thought hysteria was the bridge to modern psychiatry for

of such practicing physicians as Thomas Willis, Charles le Pois, the Thomas Sydenham already mentioned, or the dialogic inventions of Bernard Mandeville (see Appendix A, 1711).

It is folly to erect artificial disciplinary boundaries for hysteria when none existed at the time of its Renaissance genesis and seventeenth-century transformations; all the more anachronistic, given the common ground of fiction and ideology, to surmise that these discourses of hysteria belong exclusively to the province of medicine. Was the Bernard Mandeville who wrote *The Fable of the Bees* as well as the treatise on hysteria primarily a doctor or a writer? The question is as bad as the false dichotomy, doubly so because we have been brow-beaten to believe that the doctor should be the more privileged because he tells greater truths than creative or discursive writers. And who was more objective, Mandeville-the-writer or Mandeville-the-doctor? Let us be more mercilessly explicit about what is at stake in these admittedly rhetorical questions: in which academic department should Mandeville's discourses be studied today, given that fiction, ideology, and rhetoric constitute their essential discursivity? If medicine had not been privileged to the degree it has in the last three centuries, the mumbo-jumbo of hysteria as a label to describe almost every state of altered consciousness (altered from what?) would not have had to await the twentieth century for illumination of its conceptual weakness. We would understand better than we do today why psychiatrists now claim to see few 'hysterics' of *either* gender.[10] Female anatomy remains a constant, but magically, it seems, female hysterics have all but disappeared. Of course cultural circumstances and social arrangements alter, and no one wants to impoverish history by castrating its differences, or doubt that Willis and Sydenham (and later on, Breuer and Freud) did indeed treat women with hysterical symptoms, but the fictions of hysteria, especially its biological determinism and techniques of rhetorical persuasion designed primarily for positivistic male readers, also need deconstructing, demythologizing, opening up.

1.3. *Discourses of the Nerve*

We should not expect the discourses of the nerve and sensibility to be more dispassionately or objectively disposed than their predecessors or more sequentially patterned in the sense of source and influence on later discourses. What changes are the socio-cultural exigencies that

privileged. This is why seventeenth-century printed annals are so remarkably permeated with discourse debating whether men could contract the never-fatal condition. Yet Sydenham's transformation of hysteria begs for attention, especially the way he re-anatomizes it in the hypochondrium or intercostal nerve behind the lungs. By so doing, Sydenham inadvertently exposes the follies of the enterprise of hysteria among his predecessors, and displays his own phallocratic biases to the effect that if men *must be hysterical*, we doctors can demonstrate anatomically why it must be a lesser sort of hysteria. It is no wonder then why Foucault believed that after the lunatics who were sent away on ships of fools, hysterical women (which is practically to utter a tautology) had been the most marginalized group of the population of early Modern Europe (see Foucault, 1970, ch. on hysteria, *passim*).

Yet if the agenda of hysteria and its research program in Europe's medical schools sought to remythologize and mechanize the uterus in the name of suppressing women by explaining their record of failure to them, its genetic encodings also differed. Unlike melancholy, hysteria thrived on the binary opposites of male and female versus the gender-free chaos that pervades the melancholic's view of the world. Different too were hysteria's oppositions and the deterministic mythologies it sought to revive, especially the view of women as sinners, chained to their bodies, enslaved to their (suffocating) wombs, while men were 'scientifically' demonstrated not to be so, all this occurring during an alleged scientific revolution in anatomy. Hysteria will explain this legacy of female sin through the fiction of an organically 'suffocating mother'. When this explanation fails, in the eighteenth century, the discourses of sensibility, equally fictive if less organically vivid and less pictorially dramatic than hysteria's, will provide another approach equally grounded in anatomy. Small surprise then that our twentieth-century Marxist historians, like the seventeenth-century Puritans, should have been so profoundly attracted to the centuries in which hysteria flourished so exuberantly: a continuous record — so the male doctors claimed — of female sin.

The fictions of the discourse of hysteria were rhetorically charged. If melancholy had been 'anatomized' (Burton) in a genetic vehicle that sprawled, reducing itself to chaos and confusion, the discourses of hysteria had no such generic constancy. They could adapt to the Latin couplets of a sixteenth-century versifier, as well as the prose analytics

or vitalistic philosophies. They are always radical dualists or radical monists but rarely middle-liners. Their agenda often attempts to relate brain to body and thereby to construct a natural philosophy of the soul; their research programs show them as less than dispassionate, as they respond to political crisis and social exigency. Now I know that the word ideology was not coined until 1801 by Destutt de Tracy; the nerve researchers of the Enlightenment were ideological nevertheless. When we ask, what did those 'scientists' think they were doing, we eventually return to the same binary pairs of mind and body. Not all nerve research of the Enlightenment can be reduced, of course, to linguistic structures or concepts. The eighteenth-century experiments made for reflex action, for example, have little to do with language in their origination, were even more mathematical than they were linguistic, which reminds us of our enterprise here. Because we who advocate Literature and Science worry who our real audience is, we try to entice all potential audiences in the name of common assumptions; try to persuade in just the ways the linguists assure us typify ordinary speech acts, especially the common assumption that scientific models are just another set of models to describe nature's laws: ultimately neither more nor less accurate than competing models, rarely free of value or ideology when set into discursive narratives, and certainly no 'truer' than any other fictions. And yet the two aspirins that relieved my headache on the airplane yesterday are not rhetorical, ideological, value-laden or polemical aspirins until I start *talking about them*. When we return to the basic unit of the nerve — the animal spirit — we glimpse to what degree nervous discourse was itself language-bound, for the sheer number of these discursive projects is remarkable (see Appendix A for examples).

Before the nineteenth century, the 'spirit' is the sign of all the discourses of the nerve, even in those I call counter-nerve (mysticism, hermeticism, the Paracelsian and Behmenistic tradition, Stahl's animism, Swedenborg, Blake *et al.*). As William Empson long ago noticed, spirit is a word lending itself to heavy metaphorization and rhetoricity, but my point is that 'spirit' also invited ideological privileging whether in written or spoken forms. In one sense the worst accident that ever befell physiology was its intimate association with animal spirits. Yet when we recognize that *all* Enlightenment physiology was 'nervous' and necessarily grounded in the spirit, we can see why a critique of nerves developed by those who claimed to be excluded from its social place these discourses under stress, especially as they borrow from each other and forget what is inconvenient in the appropriation.

The discourse of the nerve arises when the animal spirit (already conceptualized in the sixteenth century but not yet influential) becomes the obsession of the mechanists in the middle of the seventeenth century in the aftermath of Cartesian physiology. As mechanistic and vitalistic ideologies then vied for privilege as the fundamental explanation of micro- and macrocosmic structures, and as they elicited a competing ideology of materialism during the eighteenth century, the nerve emerged as the signifier of every psychological theory of human behavior barring mystical and animistic ones; universally invoked by doctors, empiricists, moralists, ethicists, physiologists, and even diet mystics like "lettuce-and-seed Dr. Cheyne" whose *English Malady* depends entirely on the nerve for its existence. By the nineteenth century and the era of Sir Charles Bell and, much later on, Ramon y Cajal, neurophysiology is already inscribed as the fundamentum of all theories of learning and knowledge. Whether in studies of the reflex action, cerebral localization, the autonomic nervous system, or the integration of these, the nerve remains the base unit. Victor Weisskopf, the physicist, has epitomised what the nerve has meant for human destiny when viewed in geological time, and Dr. Fred Plum, the neurologist, has put his bias more succinctly if also more positivistically: "there is brain and only brain; ultimately, everything reduces to brain and can be explained by brain" (see Plum). The reduction can be interpreted in many ways and endorsed or maligned depending upon one's attitude to positivistic theories, but it has been pointed out to Plum and his colleagues that they should read philosophers like Thomas Nagel (already cited), who would assist them to understand their own modes of privileging. But the disputes between our twentieth-century philosophers and brain theorists notwithstanding, the nerve was first privileged in the Enlightenment to a degree it had certainly not been in the discourses of melancholy and hysteria, where there is rarely a sense of human beings as essentially 'nervous creatures' (in the hundreds of pages of Burton's *Anatomy* the nerves are almost never invoked).

The agenda of the discourse of the nerves is always a mediation between mind and body, less so between mind and brain; perhaps this is why its subtexts are always so hierarchical. Those who claim to pursue nerves empirically are aware of their mechanistic, materialistic,

power over him. Here then is the 'appropriation' and 'translation' of discourse through the strange tension of attraction and recoil. Not the anxiety of influence but the anxiety of *appropriation*. Not the influence of science *on* literature (those tortured arrows going in both directions) but the act of creation, in Koestler's sense, while in the imaginative heat of appropriating from another discourse, by importing and then exporting, by blurring, as would Godwin and Priestley later on, the dualisms of mind and body through the essential, irreducible life force that could regenerate him. The Tristram who self-consciously wonders if his book is "wrote against any thing" can supply one answer only: "If 'tis wrote against anything, — 'tis wrote, an' please your worships, against the spleen" (IV. xxii). Little wonder that a consumptive valetudinarian should set his Book of Life 'against' spleen, the seat of hysteria and hypochondria. The nerves, as both Sterne and Tristram knew, conveyed spleen's juices to his endangered melancholic imagination, as well as permitted the laughter — the true Shandyism — that kept him alive.

For us in this forum whose critical task is no different from all other forms of serious criticism, it is not merely the anxiety of appropriation that beckons but the larger cultural ways these doctors privileged the nerves. It was preeminently social and cultural privilege, deriving from extraordinary rearrangements then occurring in the Ancien Regime and, more specifically, between the genders. I claim no collusion between the scientists and the upper classes then, but the more I study this record, the more I grow persuaded that the agenda and research program amounted to a process resembling collusion. In the narratives and metacommentaries of the doctors (unlike the Sternes), there is usually one, unequivocal goal: the assertion of class distinction and gender boundary. Whichever route their discursive practices took, and whatever the degree to which these doctors and scientists remain inside or outside their discourses, the teleology of the agenda remains constant: to demarcate social classes and gender boundaries. This is why *female nerves* are treated as a separate category, and why the *female body* then breaks off so drastically from the male. In this sense the discourses of the nerve reveal an ideology similar to that of hysteria but with this difference: while hysteria sought to mythologize the *womb* as a means of carving out a notion of female gender, that of sensibility mythologizes, even demonologizes, the *nerves*. Both are anatomic, organic approaches yet with this distinction: if both genders have nerves, males have no wombs, females no penises. Perhaps a phallocratic

privileges. To the storytellers — the Sternes and Smolletts, the Diderots and Austens, who partook in this critique — the excesses prompted by the metaphoricity of nervous discourse were sometimes risible. But the doctors and physiologists, the Hallers and Whytts, the Cheynes and Cullens, could also be myopic to excess, or if not patently myopic (which much of the evidence affirms) then so enmeshed in the discourse as a cultural and professional 'product' that they cannot get rhetorically outside it. Enlightenment doctors, convinced of their utility to medical science, reduced human behavior to spirits, fibers, and nerves in their writings, creating a lush jungle of metaphor, which when brought under the lens of linguistic analysis proves to be a value-laden, subjective, and passionate labyrinth. Literary critics who till this era have enjoyed pinpointing the risibilities of these excesses, noticing how often animal spirits are terrifically personified, metaphorically embroidered to run on tracks and roads like carriages and trains, stop and go, are impeded by accidents, even heat up and catch cold (see Rousseau, 1976; Myers).

In Sterne's *Tristram Shandy* the animal spirits are the key to the riddle of the hero's dilemma: accounting for Tristram's messed-up personality (as the opening paragraph indicates), as well as his tragically (and by contrast at times comically) defective sense of time. Sterne derides their linguistic excesses from the start, and in one sense he deserves praise for curbing those so-called 'objective' Enlightenment doctors and physiologists. But in quite another sense, in his Rabelasian gene as it were, the Sterne who seems to be a sleepwalker in the world of Bakhtin's Rabelais, is curiously both attracted to and repelled by the animal spirits, his ambiguous state of mind a perfect specimen of the fluidity of discourse when it labors under the anxiety of appropriation found in Sterne. More specifically, when someone like Sterne is attracted by its regenerative energy to fire up his own failing imagination (Sterne was seriously ailing as he composed *Tristram Shandy*), which was all too prone to embrace the radically vitalistic notion that life exists in its most irreducible form, in the life force of the animal spirit. And Sterne, quite unlike the very different Defoe and Fielding, makes the great imaginative leap that becomes the seed of *Tristram Shandy* as he himself lingers in the throes of consumption, in the depletion of his own vital life force. Recoiling from the animal-spirit treatises he had felt impelled to reject on purely linguistic grounds, they now allure and attract him — paradoxically — by their own regenerative

transactions which make such a noise in the World, and establish Monarchies or ruin Empires, reach not so many Persons with their Influence, as do the Theories of Physiology" (Boyle, pt. ii. p. 3).

1.4. *The Discourse of Sensibility*

Sensibility is fourth, a shorthand label for irritability, excitability, the passions in all their jumble and diversity, those vital exciting powers that Hunter and Cullen and Brown will exalt and the Romantics and Germans later on, a concept, however loose and imprecise, nevertheless familiar now to literary historians and historians of science. We also know it as the prelude to Romanticism, a label, a shorthand, for the European movement diachronically subsequent to Classicism. Northrop Frye called the literature of England between approximately 1740 and 1800 "an Age of Sensibility". But the literary historians have no story to tell about the origins of sensibility, except to say that it first appears in the Restoration, in French novels like Madame de Lafayette's *Princess of Cleves* and in sentimental plays like Steele's *The Conscious Lovers*.[11] Yet imaginative literature seems to configure sensibility before science appropriates it, perhaps as imaginative literature idealizes heterosexual love. And literature will continue to transform sensibility into the nineteenth century, when novelists like Jane Austen, critical of its pretensions but appropriating its tropes altogether differently from the Sterne who by then seems to have inhabited another world, another cast of mind, turn it upside down and make it the centerpiece of *Sense and Sensibility*.

Yet it would be false to think, as some literary historians do, that sensibility was only a *literary* discourse. On the contrary, if it began there, or first appeared there in written discourse, it was soon imported into scientific writing. Albrecht von Haller, the Swiss Protestant physiologist, made sensibility the centerpiece of his physiology, claiming its complete dependence on nerves when he wrote that there can be neither sensibility nor irritability without them.[12] Haller maintained that only those parts of the body supplied with nerves possess sensibility — thereby privileging the nerves extraordinarily — while irritability is a property of the muscular fibers. It is not merely the rigid opposition Haller establishes that commands attention but the privileging of the one (sensibility and the nerves) over the other (irritability and the fibers). For his dual scheme and his sustained claim that this original difference underlies the gap between the discourses of hysteria and the nerve, but if so, it was not because the scientists were generating a neutral philosophical mechanism or vitalism, but rather as an urgent response to rearrangements of gender.

By now it is a cliche that in 1750 or 1800 many more people were writing than had been in 1550 or 1600. Those who wrote discourses of the nerve amount to at least tenfold the number as those in melancholy or hysteria, so we cannot expect the same isomorphic uniformities or similarities of view. The discourses of the nerve are more necessarily pluralized; within their ranks difference and marginality appear everywhere. Seizing 1800 as a convenient dial, a crisis of representation begins to appear among the scientists who now privilege their own discursivities above the agenda of the nerves itself, some preferring narrative in anatomy and physiology, while others endorse analytic commentary, critique, or polemic. Genetic encodings have also penetrated into every type of literary form: dialogical (as in Mandeville), discursive (as in Cheyne), poetic (as in Flemyng, the author of a long Latin epic poem published in 1747 called *Neuropathia*), satiric (as in Sterne), didactic (as in Jane Austen), fragmented (as in Coleridge), and so forth. As discursive commentaries the discourses of the nerve assert social stratifications, while their concealed hierarchy remains the supremacy of mind or body. As meta-commentaries, they range from the serious (as in political or economic treatises that metaphorically appropriate their nervous agenda, i.e. 'body politics') to the playful and the satiric, as in Sterne's "well-dress'd gentleman", who charmingly turns out to be a mere homunculus appropriated from the preformationists and epigeneticists. Always at stake in the agenda, no matter what its rhetoric, is a definition of life in terms of the dualism of mind and body, or soul and body, and a bias about what it means to be 'human' in distinction to the brute kingdom. In this sense there is no failure of nerve (*pace* Peter Gay who invokes this dramatic phrase in his study of Enlightenment culture — no pun intended). Enlightenment scientists *as well as* Enlightenment writers (it will not do to think of them separately or to read backwards, anachronistically, our own post-Kantian disciplinary boundaries) have appropriated the largest question of all, the question about life, but language has entrapped them. Small wonder then that Boyle, the great chemist but also often the most astute of commentators on the contemporary cultural scene, should write, unprovoked while in a meditative mood, that "those great

professor of medicine at Edinburgh University whose work on reflex actions remains a classic of neurophysiology, imported Hallerian sensibility to Britain, challenged it, popularized it, debated it; so did Cullen, John Brown of Brunonian medicine and, in France, virtually all the physiologists.[13] The optimistic claim was that Haller's theory of sensibility would do for the human body what Newton's calculus had done for the spheres. Today Haller's physiology is not news to the historians of science, but his charged language and social ideology is, as well as the cults of sensibility that developed around it or concomitantly with it. Like Newton, Haller knew he must expound his laws in ordinary language as well as in mathematics; he was aware how much of his success depended on the way he charged a single word: *sensibility*. The more lay culture popularized sensibility, the more scientists wanted to debate it, or transformed its original Hallerian conception to such a degree that a new version emerged, as was the case for Wordsworth at the very end of the eighteenth century. At stake here were not the metaphorical excesses and rhetorical flights of animal spirits, but the wide appeal of a doctrine claiming to be objective and unideological, yet one whose ethical implications were evident and unacceptable in several camps; moreover, a doctrine whose language (and I trust that this language is the point of interest to us) — whose language persuades and cajoles by appealing to those readers who *already* think they possess delicate nervous systems that will produce just the type of sensibility Haller describes.

How then did Hallerian sensibility progress from a physiological doctrine to the shorthand label for an entire literary movement? The route is more complicated than there is time for here and includes the tension of social arrangements then, as well as anachronistic tendencies of literary historians when they study a previous era. Even so, the scientific doctrine of sensibility was soon called upon to legitimate class distinction and gender difference. On the surface sensibility's research agenda was an early scientific positivism examining nerve tips, dermis, skin texture and color, reflex action and reflex arc in relation to norms of refined behavior, while attempting to demonstrate that acceptable social behavior is *physiologically* predetermined. Sensibility was in this approximate sense a type of eighteenth-century socio-biology. This is why fictional figures such as Sade's Justine are portrayed as having the most exquisite nerve tips in the anatomic nether region where they count for a woman of her propensities, and why — by contrast — her observation was based on years of rigorous experimentation Haller was attacked; even so, his hypothesis was more encompassing than this, was embedded in a discursive prose work every bit as polemical and ideological as Boyle's pronouncement about physiology. "Nerves", Haller wrote, "are the basis for brain and sensory impressions, for all human passion and reason, for emotion and feeling, for higher associated ideas and principles, for the thoughts of monarchs and the legislations of parliaments" (Haller, 1786, p. iv). This appears remarkably close to Victor Weisskopf and Fred Plum when they maintain, almost deterministically, that we owe virtually everything we are to our nervous apparatus.

Haller may have been satisfied to decree physiological laws, but his colleagues wanted 'sensibility' to be the basis of an entire approach to life, an ethic in itself. Haller considers the primitive sensibility of sensory impressions merely as the microcosm of sensibility's greater role in the macrocosm of human affairs, as it was displayed in human sympathy, empathy, benevolence, virtue — all the cults of sensibility in the moral realm. This was a remarkable doctrine even for an optimistic era, and it was developed by Europe's finest scientists after Newton, Boerhaave, and Leibniz. Of Haller's sensibility Condorcet wrote in the *Dictionnaire Encyclopedie*: "The work in which Haller published his discoveries [about sensibility] was the epoch of a revolution in anatomy" (Condorcet). But by 'anatomy' Condorcet signifies something other than what we do today; by anatomy he suggests a gaze, a way of seeing the entire world around him: anatomy as picture, image, vision, and — of course in view of its vestiges from the Renaissance — anatomy as dissection. Here is Charles Bonnet, the Genevese biologist claiming to be "in contact with another great man, who was soon to make in physiology the same kind of revolution that Montesquieu did in politics: I speak of Haller" (Savioz, p. 155). Again, physiology signifies something much grander than the delimited domain we mean today, its orbit and sway both temporally and spatially, in Bachelard's sense, more extensive. The catalogue can be extended. Louis Figuier, the nineteenth-century medical doctor in Paris and prolific popularizer of the entire realm of natural science, continued to write about Haller's sensibility as if it had been the scientific revolution of the Enlightenment, *the* new way of grasping interior time and space (Figuier).

Haller's hypothesis was quickly translated from Latin and German and exported from Middle Europe. Robert Whytt, the Newtonian

environment. But if the discourses of the nerve continued to purport — as our own brain theorists do today — that we *are* our neurophysiology, that we *are* synapses, that we *are* our brain — sensibility develops the same possibility while affirming class distinction and gender difference. We are dealing here with doctrines of sensibility that had *already* been in the public domain *before* the Hallers and Whytts conducted their experiments, but all along my point has been that the physiological discourse of sensibility, as distinct from other writings on sensibility, was not immune to these social factors. It assumed them, it incorporated them, it embedded them into the tropes of its discursivity. Anatomy and physiology were — to be sure — among the 'softer' sciences, as even Ramon y Cajal would concede when interrogated by his positivistic-minded colleagues in physics and mathematics. These subjects could be assimilated more easily than physics or mathematics; and sensibility was just the kind of doctrine (like Freudianism or Jungianism) that lent itself to popularization prone to distorting it from what it originally was among the physiologists. Yet even under Haller and Whytt, some of the best of its Enlightenment theorists, and among dozens of other mechanists and vitalists, materialists and anti-materialists, and all those who are not easily labeled or cubbyholed, its social biases are evident.

2. CONCLUSION: THE PRODUCT OF LITERATURE AND SCIENCE

It is time to take stock and see to what conclusions our argument has taken us. One dimension is clear even *without* conclusions: the topic sufficiently vacillates between theory and practice, between critique and metacritique, to indicate the presence of an inescapable tension. Again, this has to do with privilege of the one over the other — in this case narrative or story over critique — and the sense that theory, or metacritique, has not earned its keep, so to speak, unless validated by a prior practice.

Furthermore, two conclusions can be drawn without much reflection as they directly follow from my analysis of these four discourses. The first is that as I have tried to decipher these discourses I have continued to ask myself two main questions about what the method can yield, as well as inquire what these scientists (when they *are* scientists) thought they were doing. The more I isolated these two questions, the more I

poor Hottentot cousin Venus, in Africa, has the thickest, roughest dermis on earth, presumably with very few nerve tips in it. Yet when sensibility's ethic is exposed, as it will be around the turn of the nineteenth century, the configuration of anatomy and character alters. Blake's women (to the degree they possess this gender) — Ahania, Enitharmon, Vala — are not fibrous or nervous at all; Blake reserves nervous sensibility for his men, as if to reverse, even in gender distinctions among mythological figures, the prevalent scientific traditions of his time (see Hilton, ch. V, pp. 79—101).

The iconography of sensibility, like that of hypochondriasis, displays male sensibility as often as female. In this sense sensibility is somewhat less polarized *vis-à-vis* male and female gender. Its signs are also more varied: not merely the strength or flaccidity of the animal spirit or fiber within the nerve, but now those *in addition to* the texture of the dermis, tears of the eye, hot flush of the cheek, and natural expressions of blushing, weeping, crying, swooning, in brief all the anatomic manifestations of emotion and feeling, sympathy and empathy. A weeping or blushing male becomes the signifier of extraordinary delicacy and exquisite sensibility, in the Enlightenment thought to be quasi-hermaphroditical, in the nineteenth century evidence of dandyism or decadence.

Yet there is this difference. In the discourses we conventionally call imaginative and philosophical, the feelings and emotions are privileged over the reason and intellect as an index of sensibility, this while Haller and the physiologists are developing a so-called 'scientific' theory to substantiate the maxim that the more sensitive the nerves, the more sensitive the person. But in what sense then was sensibility an objective or dispassionate, a neutral or value-free discourse? The narrowly conceived (i.e. internalist) physiological advances that came out of sensibility were admittedly considerable and irrefutable so far as I know; but as soon as they become encoded in discourse they demonstrate another teleology. All writing, like all language, is of course metaphorical and rhetorical, and what must especially interest us here are not the general means of persuasion in a so-called 'scientific' text or discourse of sensibility, but the particular strategies used. In the immediate case, Haller's tropes are anything but neutral reports of the experiments he conducted. His discursive practices reveal what was new in the discourse of sensibility when compared to the older one of the nerves. Looking back from our gaze, we might say a confrontation was building between the forces of innate physiology versus those of

historical understanding by a type of incompleteness axiom, as well as explain my refusal to privilege metacommentary and metacritique of these discourses in their own era over ordinary commentary and ordinary critique. I must at least be willing to show why the narrative of a physiologist in 1700, for example, is neither more nor less indicative of, neither more nor less value-laden than, the species I call the discourse of hysteria, or the discourse of the nerve, than a metacritique or metacommentary on one of these narratives.

The conclusions that are less immediately obvious also derive from my approach but implicate as well, it seems to me, our recent activities in Literature and Science. They are, so to speak, the metacritique of the discourses of the nerve as it was worked up for this *particular* occasion, and they address Literature and Science in the abstract by taking stock of this historical moment. In both these senses, my approach focuses on the *peculiarity* of Literature and Science, construed either as the collective activity of a community of scholars or — alternatively — as a developing discourse that by now merits commentaries and metacritiques. By peculiarity, I mean the sense that there is something inherently different about 'performing' or 'doing' Literature and Science in an age of specialization when compared to 'doing' traditional literary criticism or even traditional philosophy. But this essential peculiarity also includes the perhaps equally alien notion — or so it must seem to many who have been trained to believe otherwise — that when scientific ideas are reduced to their verbal (i.e. literary, rhetorical, discourse) representations and encodements, scientific discourse must be treated no differently from any other type of discourse; Galileo's pronouncements deliberately cast in a fluid and vernacular Italian rather than in the accepted cadences and rhetorical Latin of the time serve as one example among many. And — what must again seem rather alien and peculiar to many contemporary scholars — the peculiarity that when science is viewed as a *discourse* rather than as a set of logico-rational inferences classified into the taxonomies we traditionally call organized academic science (biology, chemistry, physics, medicine, etc.) or accepted laboratory procedures, its 'meaning' is neither more nor less transportable or transferable than the meaning of literary and philosophical discourse. This matter of the transportability or transferability of scientific thought, in comparison with literary or philosophical thought, is not a subject about which many scholars have cared to comment. The reasons would seem to be too obvious to belabor.

had to concede how much of the *activity* of science is not linguistic, or at least attempts to break away from the constraints of language. That science today is entirely unable to break away is, I presume, one of the main rationales for the development of Literature and Science as an autonomous discipline. But in the end it would be as detrimental to our purpose (however that purpose will eventually be defined among those who work in Literature and Science), as it would be unfair to the European scientists who generated these discourses, to pretend that *all* their activity was language-bound, or that they could confront reality in linguistic categories only, as if language were the only way their brain could process a reality external to themselves.[14] Detrimental, furthermore, to pretend that there were no scientific instruments, measuring devices, mathematical procedures, laboratories, diagnoses, therapies.

The second conclusion is that in the process of self-conscious reflection on my method of privileging discourse over instrumentation and therapy, I recognize that I present this terrain in an iconoclastic and possibly even disturbing manner to professional physiologists and neurologists, for whom language often signifies impediment rather than clarification, or at least an intermediate grid whose clarificatory functions are miniscule in comparison to those of laboratory investigation and especially of predictability and falsifiability. Disturbing, moreover, to some professional historians of science and medicine who find this threshold too all-inclusive for what they consider to be 'science'. Instead of approaching this material through the conventional disciplinary filiations (i.e. by viewing these subjects biographically, within a diachronic flow of the history of physiology, or as recognized *specific problem cases* to be solved for specific moments, such as the history of melancholy in the sixteenth century, or the history of hysteria in the seventeenth); instead of the non-interdisciplinary approach of literary scholars who until recently rarely looked at the domain of science as a valid influence; instead of all these, I offer a more or less synchronic analysis that gazes from afar at a vast bulk of discursive writing without privileging *literary* discourse over *scientific* discourse, or for that matter any discourse over any other, and by construing *literary* criticism as essentially similar to every other type of incisive criticism. What I do privilege is *this particular body of discourse* within a wide array of competitors in the period of early modern Europe largely on grounds of neglect, and I must be held accountable to demonstrate the consequences of the neglect. That is, I must justify that the neglect impedes

theory, or any other type of theory here, except to suggest that discourse theory serves the purposes of, and opens up fields for, Literature and Science when placed under certain specific conditions.[17]

To conclude, if the product of Literature and Science must be a *new* theory in and of itself, then a product such as 'discourse theory' at large or the 'speech-act theory' of the linguists has greater claim to being a genuine product of Literature and Science because it was generated by humanists (that is, those trained to interpret the languages of texts) who applied a scientific approach to a hitherto humanistic enterprise, than the product of all those who have not yet discovered a new general theory as in the examples I have just provided. More tersely put, if the former, then the only way to test whether someone engages in Literature and Science is first to ask them what they are working on and then what their *new* theory is. Finally, if the former (i.e. the new theory), then we will have to address a whole range of hard questions about the professional institutionalization and implementation of our subject, which most of us have not even begun to formulate, let alone place before deans.

Robert Scholes has written in the opening sentence of his new book on semiotics that "the humanities may be defined as those disciplines primarily devoted to the study of texts" (Scholes, p. 1). What will our professional scientists think when they discover that they are 'scientists' at home but 'humanists' within the Society for Literature and Science? — an appropriation, an importation, an ambivalence, a tension, which, if anything I have been suggesting is valid, is all to the good in relation to the creative act, but which, when we have all returned to normal academic business, presents hurdles and obstacles to appointment, promotion, and tenure. And what do we say to the theorists among us who, flaunting the new discourse appropriations that disenfranchise the conventional disciplines, point out that our discussions do not concern Literature and *Science* but Literature and *Literature*, or *Rhetoric and Rhetoric*, or Tropes and *Tropes*; that is, not a conjunction at all where the force of emphasis lies on the weight of the ampersand in Literature *and* Science, but a leveling out, a reducing of all things into words, as if this newly formed society (SLS) should have been called "Society for Literature and Language", *sans* the science, as we extract the linguistic part of science and banish all the rest, as if all the human brain can do is process reality in linguistic categories.[18]

And what will our deans and provosts say when we justify funding

The consequences of the 'moment' also require standing back, so to speak, in another way, for the two derivative 'taking stocks' entail the content and assumptions of a Literature and Science talk. But *have* I just delivered a Literature and Science talk, and how can I know it if there are as yet neither regulations nor accepted conventions for the genre? I might know it by the product of the talk, and all along I have been invoking 'products' and 'yardsticks of measurement' as if I knew in advance what these products and yardsticks had been productive of. But I cannot determine whether I have fulfilled my obligation here if I do not know what 'product' Literature and Science *should* be making, in the sense that the old New Criticism was promoting a specific set of principles determined by an a priori agenda. And if I do not know at any point what that product should be, then it follows that I *cannot* know what a Literature and Science talk is.

Should this product be a new theory itself, developed because scientific approaches have been brought to bear on humanistic domains and because the scientists themselves have overlooked this domain? Or, may the product of Literature and Science be something less grandiose, such as insights into a body of discourse (as the one I have been discussing), or, even less grand as insight into a specific author or single text? If the latter (i.e. the insights) then the scientists who seek to produce this product for Literature and Science will have to initiate themselves into literary theory, for I see no way of constructing this product at any level without a modicum of theoretical grounding, in my case having profited from semiotics, speech-act theory, and discourse theory.[16] And the literary critics — at least those who work in periods before 1750 or 1800 — will have to become historians of science, or at least historians of science *manqué*, if they hope to master this body of discourse and the interpretations that have been endowed upon it over the decades.

But if the former — if the product must become a *new* theory in and of itself, which has not been developed because there have not been numbers of scholars who worked professionally in the interstices of Literature and Science, and therefore have been unable to develop a chaos theory or productivity theory, a completeness theory or paradigm theory, all these being hypothetical examples — then we are going to have to shift our notions of the diachronic aspect of the enterprise. And we will need to eliminate (to be brutally frank) three quarters of the papers usually presented in forums called Literature and Science, *mine included*, as I have certainly not developed any new neurological

explore the institutional and the academic implications of its recent blossoming. But I realize that a desideratum for this type of pluralism may be untenable in academic settings where turf counts for much, where the student's precious time is often too limited to study these ancillary matters, where the conventional university departments have the weight of decades behind them, and where for the one hundred and thirty scholars who convened in Worcester in October 1987 and who believe in building these bridges and creating these new taxonomies, there remain thousands — literally thousands — who vigorously disagree with us.

APPENDIX A

Note: The list below provides a very eclectic sampling of the types of discourse in each category, and is not intended to suggest that the list is representative of the category, let alone complete in any way.

The Discourse of Melancholy:

1586	T. Bright, *A Treatise of Melancholie*
1597	A. du Laurens, *Discourse de la conservation de la veue: des maladies melancholiques*
1600	Nicholas Breton's *Melancholike Humours*
1600	F. Platerus, various treatises on the melancholic fever
1623	E. Ferrand, *Maladie . . . melancholie erotique*
1691	T. Eggers, *A Discourse Concerning Trouble of Mind and the Disease of Melancholy*
1716	R. Baxter, *The Signs and Causes of Melancholy*
1723	W. Stukeley, *Of the Spleen*
1727	W. Harte, *Religious Melancholy*
1742	E. Synge, *Sober Thoughts for the Cure of Melancholy*
1765	A. C. Lorry, *De Melancholia et Morbis Melancholicis*

The Discourse of Hysteria:

1623 –39	Charles le Pois, various treatises on the signs in medicine and the causes of hysterical maladies
1670	T. Willis, *Affectionum quae dicuntur Hystericae et Hypochondriacae*
1682	T. Sydenham, *Epistolary Dissertation on Hysteria* (Eng. trans. 1696)
1704	M. Alberti, *De morbis imaginariis hypochondriacorum* and *Dissertatio medica de hypochondriaco-hysterico malo*
1711	B. Mandeville, *A Treatise of the Hypochondriack and Hysteric Passions*
1725	R. Blackmore, *Treatise of the Spleen and Vapours, or Hypo and Hyp*
1729	N. Robinson, *A New System of . . . Hypochondriack Melancholy*
1732	T. Dover, *Hypochondriacal and Hysterical Diseases*

for our next sabbatical leave on grounds that we humanists want to fill in the interstices between Literature and Science, forgetting the weight and ambiguity of the ampersand as well as the directions of the arrows for the moment, and overlooking the unassailable fact that the concept of science has been very greatly transformed over the centuries, as philosophers of science *manqué* or historians of science *manqué*, and that we are the *only ones competent* to accomplish this because we humanists know how to interpret texts? Will they reply, but I thought you were in the German department, or the French department? And what will be their retort to the scientists amidst us who, perhaps more naively than us but more practically, continue to ask why all these words are needed rather than well-stocked laboratories?

Finally, considering the social realities of the world we inhabit today, its corruptions and injustices, its political schisms and political polarities, what do we think we are doing in a newly founded Society for Literature and *Science* — here let us indeed emphasize the linguistic, rhetorical, and even the demagogical dimension of science — when women continue to be oppressed in the name of inferior anatomy and physiology even if we no longer call them inferior interior spaces, and when races, religions, and sexual alternatives are condemned by a new American fundamentalism, a new American moral majority, a proliferating neo-Right, which some say control billions of dollars and manipulate millions of minds, even the young American minds who sit in our college classrooms but who are not learning because a prior manipulation has prescribed that the free investigation of science is forbidden in a Christian world, and that literature is valuable only to the degree that it has been both censored and censured?

Will we reply that Literature and Science (again forgetting the very problematic ampersand and the epistemological dilemmas posed by the direction of the arrows of influence) is actually neither a new brand of humanism, as George Sarton claimed half a century ago,[19] nor a plugging of the interstices of literature and science, however construed, but more fundamentally and more practically, a 'science criticism', just as there is music criticism, art criticism, literary criticism, sports criticism; criticism almost of a journalistic type, where the author need not explain much more than the newspaper journalist?

I would like to think that Literature and Science could remain pluralistically receptive to all these approaches for the time being, an open field without fences or no-trespass signs, at least until it can

² The philosophers have written at great length about the mind-body problem, but in most instances have not done so diachronically. A specimen of the kind of question they ask is found in Matson. For a different approach see Rather.

³ See the headnote to the appendix. The taxonomical crux is genuine and cannot be dismissed: I divide this vast hulk of writing into four discourses for the purposes of diachronic convenience (i.e. arranging the four chronologically) and — more significantly — in order to demonstrate how the intrusion of 'nerves' changed each of the discourses after the mechanization of the nerves (Cartesianism) in the middle of the seventeenth century. There were theories of the nerves, of course, *before* then (Galen, Hippocrates, Vesalius, Fernel et al.) but the nerve was not considered to be the source of all life as it would be in the aftermath of Descartes. A more valid objection is why there are not *three* discourses without that of the nerves, each existing in a pre- and post-nervous (i.e. or pre- and post-ca. 1650) phase. The main reason is that the discourse of the nerve assumes a life of its own by 1650 or thereabout, and I have wanted to show its distinctness from the highly derivative but nonetheless different discourse of sensibility. But I hope to amplify this tetrapartite taxonomy more fully in *Fire in the Soul: The European Medicalization of Human Imagination*, a book in preparation.

⁴ My approach fuses this semiotic attitude with speech-act theory; the utility of both, considered individually and then in conjunction, is evident when one compares it with a more conventional historical approach.

⁵ If the question is asked: what is the competitor to the discourse of melancholy? the reply is the discourse of hysteria (not hypochondriasis, which was amalgamated with hysteria at the end of the seventeenth century). The literary tendencies (discursive practices, genetic encodings, rhetorics, ideologies, value-structures, gender differences) of each discourse are different despite certain overlaps dictated by their both being generated within the same culture; for this reason few authors who write the one discourse also write the other, as can be demonstrated bibliographically. For further evidence of the disparity of these discourses, see Jackson. Like Jackson, I have found few treatises that combine melancholy and hysteria: for hysteria read on.

⁶ I.e., as a consequence of Kant's *Critique of Judgment*, after which time it became nearly impossible for anyone philosophically aware who had read and understood Kant to construct an adequate and determinate *theory* of any discourse, let alone of several discourses, on grounds that no discursive project ever fully knows itself.

⁷ For this trope at the time, see B. Vickers; for the curriculum there, see Winstanley.

⁸ See Hodges, whose approach diminishes the role of the imagination.

⁹ See Abricosoff and Veith. There were, of course, many others, including Cesbron (1909), Bianchini (1931), Howe (1944), Brain (1963), Wajeman (1982), and Trillat (1986). Helen King has provided a revisionist argument to show that the Greeks meant something entirely different by 'suffocation of the mother' (*hysterike pnix*), and that hysteria as we know it is a modern, Renaissance invention; see her unpublished Ph.D. thesis, "Medicine and the Ancient Greeks", Cambridge University, 1986.

¹⁰ Some explanation of the reasons are found in McGrath, especially pp. 152—61.

¹¹ On the vexed matter of its origins and of the rationale for discussing origins at all in a problem of this type, see my discussion of the primary as well as secondary literature on the subject (1985).

1777   P. Pomme, *On Hysteric and Hypochondriac Diseases*
1777   J. Berkenhout, *A Treatise on Melancholy and Hypochondriacal Diseases*
1783   É. P. C. de Beauchene, *De l'Influence des affections de l'âme dans les maladies nerveuses des femmes*
1788   W. Rowley, *A Treatise on Female, Nervous, Hysterical, Hypochondriacal, Bilious Disease . . . with thoughts on madness and suicide*

Discourses of the Nerve:

1733   G. Cheyne, *The English Malady*
1738   D. Baynes (Kinneir), *A New Essay on the Nerves and the Doctrine of the Animal Spirits*
1740   M. Flemyng, *Neuropathia; sive, De Morbis hypochondriacis et hystericis*
1765   R. Whytt, *Observations on the Nature . . . of those Disorders commonly called Nervous, Hypochondriac, or Melancholic*
1768   W. A. Smith, *A Dissertation upon the Nerves*
1744   J. F. Isenflamm, *Ueber die Nerven*
1778   D. Smith, *A Treatise on Melancholy and Nervous Disorders*
1784   A. Tissot, *Traité des Nerfs*
1787   J. M. Adair, *Essays on Fashionable [Nervous] Diseases*

The Discourse of Sensibility:

1750   M. Sherlock, *Letters on Several Subjects* (Letter 3, "Of Sensibility")
1751   R. Whytt, *An Essay on the Vital and other Involuntary Motions of Animals*
1751   D. Diderot, *Elements de Physiologie*
1753   A. von Haller, *A Dissertation on the Sensible and Irritable Parts of Animals* (Latin 1753; English translation 1755)
1759   A. Smith, *Theory of Moral Sentiments*
1761   H. Smith, *Essays Physiological [the physiology of sensibility]*
1768   L. Sterne, *A Sentimental Journey*
1769   A. Tissot, *An Essay on the Diseases Incident to Literary and Sedentary Persons*
1770   P. H. D' Holbach, *Système de la Nature*
1772   M. Helvetius, "La Sensibilité Physique," in *De l'Homme*
1773   Anon., *Sensibility. A poem, written in 1773*
1775   J. P. Marat, *De l'Homme, ou des Principes et des Loix de l'Influence de l'Âme sur le Corps, et du Corps sur l'Âme*
1787   Mrs. Harriet Thomson, *Excessive Sensibility* (a novel)
1798   G. Canning, *pieces on sensibility* and *The New Morality*
1811   J. Austen, *Sense and Sensibility*

NOTES

[1] Another essay is necessary to document the point. Suffice it to say that some redress has occurred in the last decade, although the just published *Oxford Companion to the Mind* (Gregory) has no plans to add a companion volume on the body.

another paper to explain my relation to them and how their assumptions inform my practice here. Although my grasp does not lie on the cutting edge of work in these fields, I am nevertheless unembarassed to relate that I have profited from Pratt.

[17] My sense of discourse differs from Timothy Reiss in that I view the appropriation of discourse from other domains than its local one as ongoing after ca. 1700, the period where Reiss stops; nor do I believe that what Reiss calls "a discourse of modernism" is in any way identifiable, let alone completed, by 1700, requiring as it does the transformations of discourse ca. 1800.

[18] This approach may appear to be that of the trickster but the name of anything is crucial to its entire subsequent development; for the essential, even quintessential, act of naming, see Hochberg.

[19] See Sarton, who begins: "My views on the history of Science were first published in Brussels in 1912 and those on the New Humanism in 1918" (preface, p. ix). It has taken the better part of a century to absorb them.

REFERENCES

Abricosoff, G.: *L'Hysterie aux XVIIe et XVIIIe siecles*, Paris, 1897.
Bottomley, F.: *Attitudes to the Body in Western Christendom*, Lepus Books, London, 1979.
Boyle, R.: *The Usefulness of Experimental Natural Philosophy*, London, 1663.
Condorcet, M. (with d'Alembert): *Dictionnaire Encyclopedie des mathematiques*, Paris, 1789. (Quoted in Cohen, I.: *Revolution in Science*, Harvard Univ. Press, Cambridge, Mass., 1985, p. 632.)
Debus, A.: 'Science vs. Pseudo-Science: The Persistent Debate', Publications No. 1, The Morris Fishbein Center, Chicago, 1979.
Dewhurst, K.: *Dr. Thomas Sydenham (1624—1689)*, The Wellcome Institute for the History of Medicine, London, 1966.
Dewhurst, K. and Reeves, N.: *Medicine, Psychology and Literature*, Univ. of California Press, 1978.
Figuier, L. (1819—94): *Vies des savants illustres: savants du xviiie siècle*, 5 vols.; 3rd edn., Librairie Hachette, Paris, 1866—70. (Quoted in Cohen, I.: *Revolution in Science*, Harvard Univ. Press, Cambridge, Mass., 1985, p. 528.)
Foucault, M.: *Madness and Civilization*, Tavistock, London, 1970.
Foucault, M.: *The Order of Things*, Vintage, New York, 1973.
French, R.: *Robert Whytt, the Soul and Medicine*, The Wellcome Institute for the History of Medicine, London, 1969.
Frye, N.: 'Towards Defining an Age of Sensibility', *ELH* **23** (1959) pp. 144—52.
Gallagher, C. and Lacquer, T., eds.: *The Making of the Modern Body*, Univ. of California Press, Berkeley and Los Angeles, 1987.
Gregory, R., ed.: *The Oxford Companion to the Mind*, Oxford Univ. Press, Oxford, 1987.
Guthke, K.: *Haller und die Literatur*, Arbeiten aus der Niedersachsischen Staats- und Universalbibliothek Göttingen, 4, Vandenhoeck & Ruprecht, 1962.
Haller, A. von: *A Dissertation on the Sensible and Irritable Parts of Animals*, J. Nourse, London, 1755.

¹² Haller, 1755, p. 690. This is the anonymous English translation based on Simon Tissot's French translation of 1755 and now printed by O. Temkin as a supplement to the *Bulletin of the History of Medicine*, IV (1936), pp. 651—97. In the introduction Tissot claims: "The great discovery of the present age is SENSIBILITY and IRRITABILITY, described in the following treatise" (1755 ed., p. iii). For Haller (like Goethe) as a splendid example of the man of literature *and* science before Wordsworth and Coleridge conceptualized this type in the now famous preface to *Lyrical Ballads* (1800; 2nd ed. 1802), see Guthke and Rudolph.

¹³ For Haller and Whytt, see French; for Cullen, Thomson remains invaluable; for Brown and Haller, see Neubauer; for Haller and France, see Lesch; also important here is Dewhurst and Reeves.

¹⁴ This processing of all reality into linguistic categories is a crucial matter in system building, and would make an excellent topic for a future conference of SLS, especially if poststructuralist critics could talk to brain theorists and if deconstructionists could discuss the matter with those interested in artificial intelligence.

¹⁵ The historical moment constantly changes. In 1975—78 it seemed essential to assess the influence, however ambiguous it may then have seemed, of French critical theory on traditional approaches to Literature and Science; see Rousseau (1978). I surveyed the field as a literary historian, very much alive to the discussions then raging among historians and philosophers of science about the immense problem of demarcation in the two realms; see, for example, Allen G. Debus's inaugural lecture as the Morris Fishbein Professor of the History of Science and Medicine at the University of Chicago. By 1980 various groups were discussing literature and medicine in ways that seemed more vital than the older and more traditional Literature and Science; for discussion see Rousseau (1981).

My later discussions (1986 a and b) address a very different set of questions. In 1982—84 the Modern Language Association of America announced it would publish a centennial bibliography of Literature and Science and it seemed important to ask, all over again, precisely what practitioners working in Literature and Science thought they were doing; the responses, written in 1982—83, have appeared in a special issue of *University of Hartford Studies in Literature* 19 (1987), edited by G. S. Rousseau. In 1985 SLS was launched in Berkeley California and a symposium was held on different approaches to the subject; these appeared in G. S. Rousseau (ed.), *Science and the Imagination*, in *Annals of Scholarship* 4 (1986). In 1986 the University of Virginia hosted a conference on the problems of literary theory in relation to those of the philosophy of science, but science itself was hardly invoked and the talks amounted to pro or con discussions of Richard's Rorty's theoretical positions; see the selected papers of the conference which appeared as a special issue of *New Literary History* 17 (1986), no. 1 under the title "Philosophy of Science and Literary Theory". Now, in 1987—88, it seems important to annotate both the healthy existence of Literature and Science as an emerging field of inquiry (perhaps as an antidote to the positions taken in Virginia) and the diversity of approaches being adopted. No one can second guess the future, but it seems evident to me that very soon it is going to be essential to address the institutionalization of Literature and Science throughout academia.

¹⁶ I acknowledge my indebtedness while being aware of the theoretical flux of these approaches and the thorny problems they continue to pose, but I would have to write

Rudolph, G.: 'Albrecht von Haller on the Future of Science', *Journal of the History of Ideas* **35** (1974) pp. 313—22.
Sarton, G.: *The History of Science and the New Humanism*, Braziller, New York, 1956.
Savioz, R., ed.: *Memoires autobiographiques de Charles Bonnet de Genève*, Librairie Philosophique, Paris, 1948.
Scholes, R.: *Semiotics and Interpretation*, Yale Univ. Press, New Haven, 1982.
Taylor, J.: *Taylors [sic] Lamentation: The Muses morning, or funeral sonnets for the Death of John Moray*, 1618.
Thomson, J.: *An Account of the Life, Lectures, and Writings of William Cullen*, Blackwood, Edinburgh, 1832.
Tracy, D. de: *Elements d'ideologie*, 1801—18.
Turner, B.: *The Body and Society*, Blackwell, Oxford, 1984.
Veith, I.: *Hysteria: The History of a Disease*, Univ. of Chicago Press, Chicago, 1965.
Vickers, B.: *Francis Bacon and Renaissance Prose*, Cambridge Univ. Press, Cambridge, 1968.
Winstanley, D.: *Unreformed Cambridge*, Cambridge Univ. Press, Cambridge, 1935.

University of California, Los Angeles

Haller, A. von: *First Lines of Physiology*, 1751; rpt. 1786.
Hilton, N.: *Literal Imagination: Blake's Vision of Words*, Univ. of California Press, 1983.
Hochberg, H.: 'Things and Descriptions', in *Essays on Bertrand Russell* (ed. by E. Klemke), Univ. of Illinois Press, Urbana, 1970.
Hodges, D.: *Renaissance Fictions of Anatomy*, Univ. of Massachusetts Press, Amherst, 1985.
Jackson, S.: *Melancholia and Depression: From Hippocratic Times to Modern Times*, Yale Univ. Press, New Haven, 1986.
Lesch, J.: *Science and Medicine in France: the Emergence of Experimental Physiology 1790—1855*, Harvard Univ. Press, Cambridge, Mass., 1984.
Matson, R.: 'Why isn't the Mind-Body Problem Ancient?', in *Mind, Matter and Method* (ed. by P. Feyerabend and G. Maxwell), Univ. of Minnesota Press, Minneapolis, 1966, pp. 92—102.
McGrath, W.: *Freud's Discovery of Psychoanalysis: The Politics of Hysteria*, Cornell Univ. Press, Ithaca, 1986.
Myers, V.: 'Tristram and the Animal Spirits', in *Laurence Sterne* (ed. by V. Myers), Vision, London, 1985, pp. 99—114.
Nagel, T.: *The View from Nowhere*, Oxford Univ. Press, New York, 1986.
Neubauer, J.: *Bifocal Vision: Novalis' Philosophy of Nature and Disease*, Univ. of North Carolina Press, Chapel Hill, 1971.
Plum, F.: 'The Brain and the Mind: An Emergence of a New Science of Biology?', paper presented at the Cornell University Conference on "Analyzing the Inchoate: Complex Interrelations in the Humanities and the Sciences", April 16, 1987.
Porter, R.: 'Barely Touching: Social Perspectives on the Mind/Body Problem', in *The Languages of Psyche: Mind and Body in the Enlightenment* (ed. by G. Rousseau), Univ. of California Press, Berkeley and Los Angeles, forthcoming.
Pratt, M.: *Toward a Speech Act Theory of Literary Discourse*, Indiana Univ. Press, 1977.
Rather, L.: *Mind and Body in Eighteenth Century Medicine*, Wellcome Institute for the History of Medicine, London, 1965.
Reiss, T.: *The Discourse of Modernism*, Cornell Univ. Press, Ithaca, 1982.
Rousseau, G.: 'The Debate about Historical Culture and the Status of the History of 'Science', *Literature and History* **11** (1985) pp. 159—75.
Rousseau, G.: 'Literature and Medicine', *Gesnerus* **43** (1986) pp. 33—46.(a)
Rousseau, G.: 'Literature and Medicine', *Literature and Medicine* **5** (1986) pp. 152—82.(b)
Rousseau, G.: 'Literature and Medicine: The State of the Field', *Isis* **72** (1981) pp. 406—24.
Rousseau, G.: 'Literature and Science: The State of the Field', *Isis* **69** (1978) pp. 583—91.
Rousseau, G.: 'Nerves, Spirits, and Fibres', in *Studies in the Eighteenth Century: III: Papers Presented at the Third David Nichol Smith Memorial Seminar: Canberra 1973* (ed. by R. Brissenden and J. Eade), Australian National Univ. Press, Canberra, 1976, pp. 137—58.
Rousseau, G.: 'Science and the Discovery of the Imagination in Enlightened England', *Eighteenth-Century Studies* **3** (1969) pp. 108—35.

both of scientific experiment (primarily in physics) and of fiction [*Romanschreiber, Dichter*]. It exists on a higher intellectual level [*auf höherer intellektueller Stufe*] and usually precedes its concrete realization either in science or in literature. The differentiation takes place only as a derivation from the *Gedankenexperiment*. In order to become a scientific experiment, it has to respect the given structure of the factual world and correspond to it [*gute Abbilder der Tatsachen*], whereas translated into fiction it is free to combine all elements and levels of reality, and to draw conclusions from them in a way that does not correspond to its given structure.

The second step in this genetic process is therefore one of concretization. The inner 'hypothesis' on the level of thought has to be brought forth into a material reality. It has to be realized in a process that involves semiotic as well as other elements. This genetic differentiation can be represented by the following diagram:

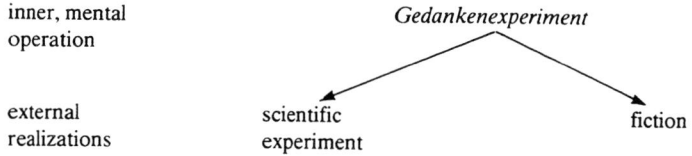

inner, mental operation — *Gedankenexperiment*

external realizations — scientific experiment — fiction

On the scientific side, the material reality is the physical world itself. Moreover, this does not mean that scientific experiment implies a passive attitude towards natural objects and processes. Quite to the contrary. Mach defines it as "*die selbsttätige Aufsuchung neuer Reaktionen, bezw. neuer Zusammenhänge derselben*" as a very active search for new aspects of the objects under study (Mach, p. 201). He is less explicit, though, regarding this second step on the side of fiction. This is not surprising since his book deals primarily with scientific research. It therefore has a built-in asymmetry with respect to our question, an asymmetry for which I shall try to compensate by equilibrating the two terms here. If the translation of the *Gedankenexperiment* into scientific experiment requires a material setting under determined and controlled conditions, its realization as fiction takes place in the semiotic materiality of language.

This can be illustrated by a certain kind of entry in Lichtenberg's *Scribbler's Diary* [*Sudelbuchnotizen*]:

WALTER MOSER

EXPERIMENT AND FICTION

A whole series of preliminary remarks would be necessary to explain how broad the question of experiment and fiction is and how little of it can be covered in this essay. Instead, I prefer to accept the fragmentary and somewhat incomplete nature of this text which — compared to the vast field of research it opens — can only have an introductory status.

1. THE FULL CIRCLE

Let us first adopt a genetic perspective and try to find an answer to the question: how does the difference, the conceptual dichotomy — if it is one — between experiment and fiction come into being? This is the answer given by Ernst Mach in his *Knowledge and Error* [*Erkenntnis und Irrtum*]:

> Besides physical experiments there are others that are extensively used at a higher intellectual level, namely thought experiments. The planner, the builder of castles in the air, the novelist, the author of social and technological utopias is experimenting with thoughts; so, too, is the hardheaded merchant, the serious inventor and the enquirer. All of them imagine conditions, and connect with them their expectations and surmise of certain consequences: they gain a thought experience. However, while the former combine in phantasy certain conditions that never occur together in reality, or imagine these conditions accompanied by consequences that are not connected with them, the latter, whose ideas are good representations of the facts, will keep fairly close to reality in their thinking. Indeed, it is the more or less non-arbirary representation of facts in our ideas that makes thought experiments possible. For we can find in memory details that we failed to notice when directly observing the facts. Just as in memory we may discover a trait that suddenly reveals a man's character hitherto misread, so memory offers new and so far unnoticed features of physical facts and helps us to new discoveries.
> Our ideas are more readily to hand than physical facts: thought experiments cost less, as it were. It is thus small wonder that thought experiment often precedes and prepares physical experiments.[1]

In this text Mach makes it quite clear that he sees the thought experiment [*Gedankenexperiment*] as the common origin, or inner prototype,

processes has been located in the hypothesis or — as we have seen in Mach — in the *Gedankenexperiment*. A century before Mach, Lavoisier saw the scientific hypothesis as a very risky and yet necessary scientific operation. It is necessary in the process of acquiring new scientific knowledge, and thus in extending the limits between the known and the unknown. It is risky, however, because it represents the moment when the scientist leaves the firm ground of referentially verifiable truth and builds a bridge to the unknown.[5] Initially, only one end of this bridge relies on facts, while the other reaches out for new grounds of knowledge. It anticipates possible facts. Thus, for a short moment, the scientist uses language in what Lavoisier deems to be a scientifically unproper way: he speaks without referring to given facts; he departs from the obligation to moor his discourse in factuality, which is said to be the ontological basis for scientific truth. In this specific moment, the scientific discourse adopts the (onto)logical status of fiction (cf. Searle). All the methodological precautions to 'contain' the risk of hypothesis cannot undo this discursive 'heresy'.

Hypothesis represents the moment in scientific activity when the facts are still silent. Lavoisier has clearly indicated that the scientist's *horror vacui* would have to be translated, in terms of his discursive practice, into what he calls "the silence of the facts".[6] Unfortunately for Lavoisier, language can be used without reference to facts: language can also refer to illusionary facts. Language can anticipate facts. The very possibility of such language uses constitutes the foundation of fictional as well as hypothetical discourse. At the same time it is the nightmare of the empirical scientist and the capital sin of scientific discourse: speaking without precise reference to given, assured facts. Lavoisier is very outspoken on this in his "Preliminary Discourse" ["Discours préliminaire"] to the *Elementary Treatise on Chemistry* [*Traité élémentaire de chimie*] (pp. 178—94).

The question of how to control hypothesis in the production of scientific knowledge, and especially with respect to experimental activity, can then be expanded into the question of how to keep fictional elements out of the realm of science. This question has been raised over and over again. Let us consider a few examples.

In the eighteenth century, the label 'experimental philosophy' was still very much used to identify a certain scientific paradigm, although compared to its seventeenth-century predecessor (especially in the Royal Society of London), it had undergone changes. In the articles

A speech and sound device such that when one speaks something into it in a foreign language, it emanates from another hole translated into German.

If a shaft were sunk through the center of the earth, one would be able to leap into it without difficulty; if one weren't killed by the air at the center of the earth, one would attain a velocity such that one would fall to the other end of the shaft and arrive very gently.[2]

For obvious technical reasons, these two *Gedankenexperimente* by Lichtenberg could not be carried out as actual, scientific experiments.[3] Nonetheless, they are already realized as entries into his *Scribbler's Diary*, and therefore have the status of fictional texts. They are 'fictional experiments', and represent a category, or even a genre, widely practiced in the eighteenth century, as we shall see later on.

Back to Mach. In his view, both the experiment and the fiction are "natural continuations of the *Gedankenexperiment*" (Mach, p. 201). In both cases, an inner and immaterial stage of the operation is said to precede its outer, material concretizations. This entails an argument which is phrased in economic terms: "*Wir experimentieren mit den Gedanken sozusagen mit geringeren Kosten*" (Mach, p. 187). To experiment on the level of pure thinking is less expensive, or — to adopt Mach's own metaphor — more cost-effective than to manipulate physical facts and data in science (or even to produce a written fictional text). Since on a very general level Mach sees scientific activity as the most economical way for man to master his natural environment, it is not surprising that he is somewhat prejudiced in favor of the cost-saving *Gedankenexperiment*.

Moreover, what seems to be the decisive difference between experiment and fiction lies in the relationship of the scientist and the artist to a given reality. The scientist is bound to depict the given natural world *as it is*, even in those features that are not directly accessible to the human senses. He has to represent (describe, depict, explain) a given real object, or at least make its factual manipulation possible through a precise code of correspondence. The artist, more specifically the author of fiction, has as a matter of principle the freedom to invent any kind of reality, or at least to combine elements of a given reality as he pleases.[4] By virtue of a verbal realization, he can propose all kinds of realities, or even transpose the given physical or social reality into an open series of possible realities.

Traditionally, the intersection between experimental and fictional

experimentation. This transfer from science to literature usually appears in a much more positive light than the intrusion of fictional elements into science. It takes place on different levels of the literary text and is realized in a great variety of forms. Let us consider some examples.

(1) Fiction can integrate the scientific experiment as a story to be told. The experimental activity is then part of the narrated content. This is the case in many texts of science fiction, such as Mary Shelley's *Frankenstein* and Villiers de l'Isle-Adam's *L'Eve future*. Goethe's *Wahlverwandtschaften* [*Elective Affinities*] also belongs in this category, although it represents a much more complex case: there the transfer of a chemical experiment into the social and psychological sphere reveals itself to be exceedingly problematical, and even catastrophic, since it brings about the self-destruction of two out of the four main characters.[11] This negative outcome places the scientific experiment's underlying rationale in a critical light. The novel calls forth an epistemological retroaction upon a certain scientific model — at least for the reader who wants to read it on this level.

(2) In certain instances, and at specific moments, fiction has itself adopted a scientific method, either by borrowing it as such from a scientific discipline or by developing in its own right the equivalent of an experimental method. The naturalistic novel in general, and Zola's in particular, is probably the best example of this category; all the more so since Zola is the author of a manifesto entitled *The Experimental Novel* [*Le roman expérimental*], which proposes to transfer Claude Bernard's experimental method from physiology to the writing of novels. Fiction thus should become an experimental instrument for developing new psychological as well as sociological knowledge which should in turn enhance the moral and social status of mankind by allowing for enlightened interventions to improve the health of the 'social body'. While claiming an experimental, scientific status in its dealings with the subject matter, the naturalistic novel remains quite traditional in its formal features and norms.

(3) Moreover, the experimentation can be transferred to the level of form and language itself. In this case it affects the very medium of literature, the rules of its own construction. This is what has been developed in the twentieth century under the label 'experimental literature' (cf. Hartung). As 'concrete poetry', this mode of literary production is closely linked to the avant-grade movements and might already have outlived itself. As examples I would like to mention Musil's

"experiment" ["Expérience"] and "experimental philosophy" ["Philosophie expérimentale"] of the *Encyclopédie* we find characteristic descriptions of experimental activity as conceived in the eighteenth century. It is defined as a very dynamic activity that consists of

> ... attempting to penetrate [nature] more profoundly, to steal what it hides, to create in some way, by different combinations of bodies, new phenomena to study them: finally, it does not limit itself to studying nature, but it examines and presses nature.[7]

In many instances, this activity is also defined negatively, in terms of what should be avoided:

> The most discerning speculations and the profoundest meditations are only vain fancies if they are not founded on exact experiments.[8]
>
> Descartes, and Bacon himself, notwithstanding how much philosophy is obliged to them, would have been more useful to it if they had been more practical and less theoretical physicists; but the idle pleasure of meditation and even of conjecture carries away great minds.[9]
>
> ... the knowledge of the hidden facts that one ascertains by seeing them, and not the novel of supposed facts that one guesses well or poorly, without seeking or seeing them.[10]

The negative attitude of not following the experimental method is denounced very explicitly: "speculation", "vain fancies", "idle pleasure of meditation", "novel of supposed facts". In these different formulations, the text systematically identifies the same disorder and deformation of scientific discourse: dissociating words from facts; allowing them to be used without referential control; giving them theoretical, conjectural, and imaginative autonomy. This whole attitude amounts to producing a "novel of supposed facts" and ought to be rejected. Here again, as we have already seen in Lavoisier's text, the features of fictional discourse are identified as what should be overcome or at least avoided. They have found their way into science, or rather must be kept from doing so. It becomes evident that fiction, from the scientific point of view, is seen as a mostly unwelcome intruder. Since science still is closely related to philosophy, as becomes explicit in the phrase 'experimental philosophy', it is not surprising to observe that philosophers fight the same battle against the same intruder. Kant, for instance, likewise rejects what he calls "philosophical novel" [*philosophischer Roman*] or *Robinsonade* in philosophical discourse.

This unruly trespassing of fiction on the scientific realm of experiment has its symmetrical counterpart. Fiction can play host to all kinds of elements and methods that belong to, or are derived from scientific

tion in order to indicate the existence of an invisible historical threshold. It seems to me that a decisive change took place in the relation of experiment and fiction between 1750 and 1800. Before that change, the two notions and the corresponding practices were not neatly separated and could, under certain conditions, even be combined. They both enjoyed something like equal treatment. After the change, as I have already pointed out, it became important to keep them apart, and fiction was maneuvered into a position of inferiority as compared to experiment. At least this is how this development is depicted from the point of view of science. Needless to say, the initiative for this change originated there.

This can be illustrated by a few examples, although a complete analysis would require more space than is available here. In French texts produced around 1750 it is fairly common to find a textual procedure which I propose to call *fictional experimentation*. It consists of the staging of an experiment whose realization is restricted to its description and narration in language. It is like an accurate verbal account of a Machian *Gedankenexperiment*.

In 1754 Condillac publishes his *Traité des sensations*. This treatise on the genesis of human faculties and understanding is entirely based on the hypothetical assumption that the philosopher can take an experimental stance and observe a statue that responds to his commands like a philosophical robot. The function of this Pygmaleonesque fiction goes beyond that of a rhetorical device, because it has decisive heuristic importance. In addition, it imposes upon the philosophical discourse the structure of a genetic narration.

Yet, from a pragmatic point of view, the 'experiment' only works if the reader agrees to entertain the fiction. Condillac is very much aware of this condition: "I let you know therefore that it is very important to put oneself exactly in the place of the statue that we are going to observe".[13] Having thus established the conditions under which the experiment has to take place, he can begin:

> Our statue's perceptions, being limited to the sense of smell, can only extend themselves to odors. It can no more have ideas of extension, of form, or of anything outside itself, or outside its sensations, than it can of color, sound or taste.[14]

From this explicit arrangement of the fictional experiment it becomes quite clear that the experimenting philosopher wishes to limit the

'essayism',[12] and more recently Arno Schmidt's and Heissenbüttel's fiction. We have to be aware that these are but a few of the many different ways of importing experimental activity into fiction, or at least to reactivate the scientific concept of 'experiment' in aesthetic production.

Having started out with the *Gedankenexperiment* which Mach presents as the ancestor of both scientific experiment and fiction, we proceeded to explore the difference between scientific experimentation and literary fiction proper, and finally observed various ways in which the two notions and practices can interpenetrate. Although we have seen that the movement in each direction receives a very different evaluation, we have now come full circle, moving from a common origin to the existence and consolidation of two separate and potentially antagonistic activities, and finally back and forth across the dividing line. This double movement reaffirms the inner connection between the two activities and, in a dynamic rather than a genetic way, reestablishes their common ground, if not their interchangeability, despite the historical process of their institutional differentiation in science and literature. The diagram can now be completed in the following fashion:

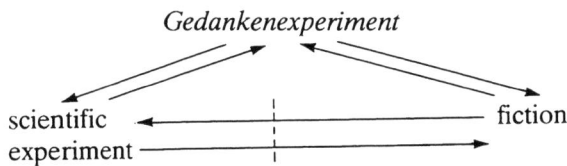

2. THE HISTORICAL DIMENSION

Although I have mentioned examples from three different centuries, thus indicating that we are dealing here with a long-term phenomenon, I have yet to articulate the historical dimension. How can this important dimension be integrated into the argument? Does this ambidirectional interaction between experiment and fiction apply only to a certain historical situation? And more precisely: is there a *terminus post quem* for its application?

Having to deal with long-term developments, we should not expect to be able to identify this *terminus post quem* as the month or the year when 'it' happened. We should rather engage in a process of interpolato give the full account of the natural history of man, which must include his origin. He explicitly formulates the questions to which only the fictional experimentation of the statue can produce the answers:

> How do our first ideas reach our soul? Have we not forgotten all that happened in the darkness of our childhood? How will we remember the first trace of our thoughts? Is there not even a foolhardiness in wanting to return all the way there? If the thing were less important, one would be correct in blaming us; but it is possibly more worthy than any other of occupying us: does one not know that one must make every effort if one wants to attain a great goal?[18]

All these questions, as well as their urgent and even dramatic undertone, confirm my decision to recognize in this use of the statue an epistemic strategy rather than a rhetorical figure. The program of historical discourse obliges Buffon to tell the beginning of each natural object. But how can man himself acquire knowledge about his own origin? Through his own questions, Buffon admits that the discourse of natural history meets its own epistemic *aporia* in the question of the origin: on the one hand, this discourse pretends to be based on factual evidence, while on the other the unavoidable question of origin has, strictly speaking, no factual answer. Hence the necessity of resorting to a discursive device. Around 1750, philosophical (Condillac) and pre-scientific (Buffon) discourse had a choice between two alternatives: either adopt the biblical myth of the origin, or make up an as-if account in which, however, the awareness of its fictional status is inscribed.

The question of the origin is then the one that brings about this epistemic stratagem. It is the unanswerable question par excellence. In Marivaux, the question of the origin of human society is raised yet again:

> P. — ... it's nature itself that we will examine; only it will settle the question once and for all ...
> To know for certain whether the first inconstancy or the first infidelity came from a man, as you maintain, and I do as well, it would the necessary to have been present at the beginning of the world and of society.
> H. — Without a doubt, but we were not there.
> P. — We will be there; yes, the men and the women of that time, the world and its first loves will reappear before our eyes as they were ... .[19]

The two persons engaged in this worldly, and at the same time philosophical discussion are two characters from Marivaux's play *The Dispute* [*La Dispute*]. Since they are unable to resolve their dispute by

variables of his operation, or at least to introduce them one by one, and at every step to observe 'what happens'. He proceeds analytically, and by doing so controls and determines the progressive unfolding of the experiment. *Incipit experimentum*: "If we present it a rose, in relation to us it will be a statue that smells a rose; but in relation to itself it will be only the odor of this flower . . ."[15] From here on, what will happen is quite predictable for readers who are acquainted with sensualistic philosophy. What is surprising, though, is the fact that in the middle of the eighteenth century, the philosophical discourse could still make use of such fictional experimentation, without losing its credibility.

This discursive device seems to be fully accepted. Yet it is more than a rhetorical ornament merely added to the truth-seeking function of philosophy. Condillac sets it up very carefully, as the scientist does in preparing an experiment, but once established it fully determines the mode of discourse that becomes the narrative account of a genetic process. Moreover, Condillac relies fully on this mode of discourse.

In view of its importance I propose to call such fictional experimentation an epistemic stratagem. The need for such a stratagem becomes quite evident in Buffon's *Histoire Naturelle* where the same device of the human statue had already been used five years earlier:

> I therefore imagine a man such as we can believe the first man was at the moment of creation, that is to say a man whose body and organs would be perfectly formed but who would awake completely new to himself and all that surrounds him. What would be his first movements, his first sensations, his first judgements? If this man wanted to recount his first thoughts, what would he have to say to us? What would this history be? I cannot dispense with making him speak for himself, in order to convey the facts more sensitively. This philosophical story, which will be short, will not be a useless digression.[16]

Admittedly, Buffon has taken more precautions than Condillac to integrate this fiction into his historical account of nature and to make it acceptable in a discourse that has to produce the truth about natural objects.[17] He is aware of the problematic referential status of such an account in natural history. He therefore organizes and realizes it very explicitly. He introduces it with the formula "I imagine that . . ." [*j'imagine que* . . .], he puts the verb in the conditional form and mode, and he identifies the genre he is using it in as a "philosophical story" [*récit philosophique*].

But he also admits that he cannot do without this fiction if he wants

few. All these are problems whose solutions cannot be empirical. The fictional experiment offers a substitute for empirical data. In this manner the question can be dealt with *as if* it were an empirical one. Around 1750, this was a perfectly acceptable fiction of an experimental method, or an experiment that was carried out as a fiction.

Less than fifty years later, this stratagem was no longer accepted, neither in philosophy (as we have seen in Kant's rejection of the *philosophischer Roman*) nor in 'mainstream' science (as we have seen in Lavoisier's dismissal of fictional elements in scientific activities). A vague consciousness of the difference between the two procedures (fiction and experiment) that manifested itself in Buffon's text has grown into the awareness of their incompatibility. This separation becomes a dogma especially within science, explicitly stated by authors who claim to legislate on modern scientific methods. They expel fictional elements from science in the process of giving science the serious and useful status of an activity that produces truth-claiming statements about natural objects and allows for the manipulation of those objects. In their view, fiction belongs in its proper place, and that is in literature.

This separation is made even deeper by an additional axiological distinction which makes experimentation appear as a worthwhile, serious and useful activity, whereas fiction is seen as an inferior, less important and less useful activity. This, of course, is the point of view of science. But it became a widely accepted, and therefore a dominant attitude in nineteenth-century society.

It is not surprising that, under the dominance of this value system, fiction, and literature in general, strives for public recognition by claiming to become itself experimental according to scientific standards. This might be seen as a late reaction to its having been separated and even expelled from the realm of scientific activity. At the same time literature reacts in this way to the important development of science in the second part of the nineteenth century. To put it in Karl Eibl's terms, literature is not satisfied any more with its subsidiary function with respect to science and moves into a position where it can adopt a complementary function and even compete with science in the exploration of human behavior.[22]

This is clearly the case in Zola's programmatic text *The Experimental Novel* [*Le roman expérimental*] (1880) and can still be observed after the turn of the century in Bertold Brecht's "On Experimental Theater" ["Über experimentelles Theater"], written in 1939. In doing

pure reasoning with arguments, they decide to resort to an experiment, set up, by chance, by the father of one of the dialoguing partners:

P. — ... eighteen or nineteen years ago, today's dispute arose in my father's court ... My father, who was by nature quite a philosopher, ... decided to resolve this question once and for all. Four children from the cradle, two of your sex and two of ours, were brought to the forest where he had built a house expressly for them ... They will then be permitted to leave their enclosure and to meet each other for the first time ... one can see the interaction they will have as the first age of the world....[20]

In this setting of the experimental stage — in the most literal sense of the word — we do not find the slightest trace of Buffon's precautions.[21] This is due to the fact that the scene is part of a discourse that is admittedly fictional at the outset: literature. This makes it possible for Marivaux to set up the fictional experimentation much more naively as compared to what we have observed in Buffon's text.

From a logical point of view, Marivaux runs into various problems with his experiment. The artificial state of nature in which the four experimental subjects have been living up to the crucial moment of the experiment is, of course, highly problematical. For instance, the four have been taken care of — if not educated — by two servants, from whom they have acquired the use of language. It is evident, then, that they have already undergone a certain socialization and that this diminishes the experimental value of the whole undertaking. But this is not the question here. We are rather focusing on the fact that, again, the question of the origin — because it can have no empirical answer — is put in the hybrid form of fictional experimentation and developed hypothetically. This form can be found on both sides of what later becomes the dividing line between the discourse of fiction (literature) on the one hand, and the discourse of empirical experimentation (science) on the other.

Be it naive or not in its textual execution, the use of fictional experimentation is always the same, across the boundaries of different types of discourse: philosophy, natural history (one of the predecessors of contemporary science) and literature. It always has to produce an answer to the question of the origin. The eighteenth century was very fond of this question and carried it over from what we today call science to philosophy and from philosophy to literature — and then back again. There was the question posed by Molyneux, the problem of the founding, the problem of the origin of language, to mention just a

motivated, of course, by the striving for a better society which sounds like an echo of the eighteenth-century theme of the 'perfectibility of man'.

In this project of an experimental novel, literature thus competes with science by adopting its method and its social legitimation: practical usefulness [*utilité pratique*]. Although naturalism was a rather short-lived literary movement, this ideological justification of literature has not disappeared with it. In 1939, Bertold Brecht uses very similar formulations to justify the use of an experimental method in drama:

... be able to intervene profoundly in social development.

... the attitude ... that seeks changes; that aims to master nature.

... the viewer ... as the great modifier, who is able to intervene in the processes of nature and the processes of society; who no longer merely accepts the world as it is, but rather masters it.[24]

The phrases we find in both Zola's and Brecht's text are surprisingly similar and contain the same kind of modern attitude toward knowing and changing society. They propose an analogous transposition of the pattern of "available knowledge" [*Verfügungswissen*] (Habermas, 1973) from science to fiction or drama. That is, they use the literary discourse to analyse complex social processes, with the final intention that the insights into the mechanisms of society thus offered contribute to bring about changes.[25] According to this practice of literature, the scientist and the literary author find themselves in a position of complicity with respect to the task of exploring and shaping the future of human society.

There is a difference, though, between Zola's and Brecht's project; it concerns, respectively, the role of the reader and of the audience. While Zola focuses on the task and the activity of the artist as an experimenter and gives him the key role, Brecht is mainly interested in transmitting the social knowledge produced in his plays to the spectator. It is essential for him to make the spectator conscious of social processes and to maneuver him into the role of the active subject of social change. It is the spectator, in the final instance, who has to take on the Promethean task of carrying out social change. In doing so Brecht orients the experimental aspect of drama towards the reactualization of the emancipatory interest of knowledge. There is nothing in Zola's project that would articulate this interest and allow the naturalist movement to resist its integration into the ideological horizon of social technology. If the knowledge about society, as produced in literature

so, literary authors and theoreticians recognize the scientific method as a model to be imitated and matched. Thus, in relation to science, literature enters a process of emancipation. Fiction claims to be as experimental as science, and to stand in the same relationship to society as science to natural objects. Yet this emancipation of literature in terms of scientific seriousness as well as practical usefulness turns out to be very problematical, because it implies the recognition of a certain scientific paradigm as the sole way of acquiring socially valid knowledge. While adopting experimental methods according to standards developed in natural sciences, literature submits itself to the leading role of science as the discipline that offers a general cognitive model.

It is not suprising then that literature also adopts from science what I would call the *ethos of modernity*. This ethos consists in the desire to go beyond pure cognition, to master and to manipulate the objects of knowledge. We find this ethos manifested in many phrases used by Zola:

We should modify nature, without leaving nature.

. . . institute an experiment, to analyse the facts and to bring them under control!

We will see that one can act on the social milieu, while acting on the phenomena which will have been brought under control by man

all-powerful man will have mastered nature

the practical usefulness and high moral standard of our naturalist works: . . . it will only be necessary to act on individuals and on milieus, if one wants to arrive at the better social condition . . .[23]

These are just a few among many textual elements in Zola's manifesto that can be related to the modern ideology of scientific usefulness. In all these formulations, the knowledge produced in the novel about social and psychological mechanisms is intimately linked to the possibility of its practical application. At the very outset, the production of this knowledge is guided by the interest in bringing its object under control. Thus, Zola sees the cognitive aspect of novel writing as inseparable from a manipulatory aspect that could be activated in a later stage on the basis of the scientific results derived from novel writing. The frequent expression "to bring something under control" [*s'en rendre maître*] covers both aspects and reveals a fantasm of mastering the world, both natural and social. Zola gives examples of social interventions that could eventually be brought about by novel writing, such as the elimination of criminality. Such interventions are

activity was no longer directed exclusively at referentially handled objects: man, society, etc. — that is, at the contents represented in fiction. It started to include the very medium of literature as well: language and aesthetic forms. Although the notions 'experiment' and 'experimental' still have scientific connotations, this new literary practice has much more of an autonomous status, since it is applied to language as its main object. Thus experimental literature — particularly concrete poetry — proposes a radical kind of experimentation *in* language *on* language.

This literary experimentation has the potential of throwing a critical light on science itself, insofar as science also is based on language, and particularly on the referential use of language. It therefore becomes possible for the experiment in fiction to have a retroactive effect on science in general. This effect is a critical one. Moreover, its impact is stronger when the literary author combines purely linguistic with thematic experimentation. This is already the case in Goethe's *Wahlverwandtschaften* [*Elective Affinities*]. More recently one finds this critical retroaction of literary experimentation on science in Musil's *The Man Without Qualities* [*Der Mann ohne Eigenschaften*]. In both works, experimental activity is both a thematic representation and a literary method, but it also has the effect of problematizing a dominant scientific paradigm.

NOTES

[1] "Außer dem physischen Experiment gibt es noch ein anderes, welches auf höherer Stufe in ausgedehntem Maße geübt wird — das Gedankenexperiment. Der Projektenmacher, der Erbauer von Luftschlössern, der Romanschreiber, der Dichter sozialer oder technischer Utopien experimentiert in Gedanken. Aber auch der solide Kaufmann, der ernste Erfinder oder Forscher tut dasselbe. Alle stellen sich Umstände vor, und knüpfen an diese Vorstellung die Erwartung, Vermutung gewisser Folgen; sie machen eine Gedankenerfahrung. Während aber die ersteren in der Phantasie Umstände kombinieren, die in Wirklichkeit nicht zusammentreffen, oder diese Umstände von Folgen begleitet denken, welche nicht an dieselben gebunden sind, werden letztere, deren Vorstellungen gute Abbilder der Tatsachen sind, in ihrem Denken der Wirklichkeit sehr nahe bleiben . . .

Unsere Vorstellungen haben wir leichter und bequemer zur Hand, als die physikalischen Tatsachen. Wir experimentieren mit den Gedanken sozusagen mit geringeren Kosten. So dürfen wir uns also nicht wundern, daß das Gedankenexperiment vielfach dem physischen Experiment vorausgeht, und dasselbe vorbereitet" (Mach, 1905, pp. 186—7).

[2] "Ein Sprech- und Schallwerk, wenn man etwas in einer fremden Sprache hinein

conceived as experimentation, should enable mankind to improve society, Brecht wants to make sure that the reader/spectator himself is instituted as the subject of this double process of knowing-and-changing. In this sense Brecht carries on what Habermas (1981) has recently called "the still unaccomplished project of modernity".

3. CONCLUSION AND OUTLOOK

The exploration of the field of interaction between experiment and fiction — even in this sketchy presentation — is far from being completed. Only two aspects — one genetic and the other historical — have been dealt with here.

The genetic approach to the whole issue has primarily a heuristic function. It has the advantage of identifying, prior to any differentiation, the common ground of experiment and fiction. It is only on the basis of such a common ground, for instance a genetic ancestor called *Gedankenexperiment*, that we can understand the complex circulation of discursive elements which cross in both directions whatever dividing lines have been established.

These dividing lines are historical constructs and therefore subject to change. Yet it is important to recognize that the existence of formal and functional differentiation is the precondition for the existence of interactions. The appropriation of 'experiment' by the scientific discourse and institution entails, as a corollary, a strengthening of the link between fiction and literature. The interactions that were thus made possible at the end of the eighteenth century were readily reduced to a one-way transpostition from science to literature. The movement in the other direction was perceived as a transgression, as disorderly discursive conduct. Within the ideological framework of modernity, science acquires a hierarchically superior position with respect to literature and can therefore impose its own practice as a general model of cognition. Henceforth we observe the desire, among literary authors, to make their art "experimental" and to have it acquire the legitimation of 'social usefulness'.

Two more chapters would be needed to complete the full story of the interactions between experiment and fiction: one on avant-garde experimental literature in the twentieth century and another on the critical retroaction of fiction on science.

At the beginning of the twentieth century, a decisive change took place within the denomination 'experimental literature'. Experimental

[14] "Les connaissances de notre statue bornée au sens de l'odorat, ne peuvent s'étendre qu'à des odeurs. Elle ne peut pas plus avoir les idées d'étendue, de figure, ni de rien qui soit hors d'elle, ou hors de ses sensations, que celles de couleur, de son, de saveur" (Condillac, p. 224).

[15] "Si nous lui présentons une rose, elle sera par rapport à nous une statue qui sent une rose; mais par rapport à elle, elle ne sera que l'odeur même de cette fleur..." (Condillac, p. 224).

[16] "J'imagine donc un homme tel qu'on peut croire qu'il était le premier homme au moment de la création, c'est-à-dire un homme dont le corps et les organes seroient parfaitement formés mais qui s'éveillerait tout neuf pour lui-même et pour tout ce qui l'environne. Quels seraient ses premiers mouvements, ses premières sensations, ses premiers jugements? Si cet homme vouloit nous faire l'histoire de ses premières pensées, qu'aurait-il à nous dire? Quelle serait cette histoire? Je ne puis me dispenser de le faire parler lui-même, afin d'en rendre les faits plus sensibles. Ce récit philosophique, qui sera court, ne sera pas une digression inutile" (Buffon, p. 214).

[17] In his major work *Natural History* [*Histoire Naturelle*] man becomes one of the objects of natural history.

[18] "Comment nos premières connaissances arrivent-elles à notre âme? N'avons-nous pas oublié tout ce qui s'est passé dans les ténèbres de notre enfance? Comment retrouverons-nous la première trace de nos pensées? N'y a-t-il pas même de la témérité à vouloir remonter jusque là? Si la chose était moins importante, on aurait raison de nous blâmer; mais elle est peut-être, plus que toute autre, digne de nous occuper: et ne sait-on pas qu'on doit faire des efforts toutes les fois qu'on veut atteindre à quelque grand objet?" (Buffon, p. 214).

[19] "P. — c'est la nature elle-même que nous allons interroger; il n'y a qu'elle qui puisse décider la question sans réplique...
Pour bien savoir si la première inconstance ou la première infidélité est venue d'un homme, comme vous le prétendez, et moi aussi, il faudrait avoir assisté au commencement du monde et de la société.
H. — Sans doute, mais nous n'y étions pas.
P. — Nous allons y être; oui, les hommes et les femmes de ce temps-là, le monde et ses premières amours vont réapparaître à nos yeux tels qu'ils étaient" (Marivaux, p. 1348).

[20] "... il y a dix-huit ou dix-neuf ans que la dispute d'aujourd'hui s'éleva à la cour de mon père... Mon père, naturellement assez philosophe... résolut de savoir à quoi s'en tenir, par une épreuve qui ne laissât rien à désirer, quatre enfants au berceau, deux de votre sexe et deux du nôtre, furent portés dans la forêt où il avait fait bâtir une maison exprès pour eux.... On va donc pour la première fois leur laisser la liberté de sortir de leur enceinte et de se connaître... on peut regarder le commerce qu'ils vont avoir ensemble comme le premier âge du monde" (Marivaux, p. 1349).

[21] This setting of an experimental stage within the dramatic play confirms the structure of the *double registre* Jean Rousset has brought out in his essay "Marivaux ou La structure du double registre".

[22] One has to mention here the rather short-lived, yet no less important, romantic alternative. Romanticism brought about a different relation between science and literature, a relation that reversed the dominance of science over literature. This is quite explicit in authors like Novalis and Coleridge. Moreover, their proposal to integrate

redet, so schallt es zu einem andern Loch ins Deutsche übersetzt heraus" (Schöne, p. 60, J 1659).

"Wenn ein Schacht durch den Mittelpunkt der Erde getrieben würde, so würde man ohne Hindernis hinein springen können, wenn sonst die Luft einen nicht tödete zum Mittelpunkt der Erde, würde man eine Geschwindigkeit haben mit der man wieder bis an die andere Öffnung des Schachts fiele und ganz gemächlich ankäme" (Schöne, p. 85, A 200).

[3] One of them, automatic or machine translation, is now in the process of being realized — at least partially — with the help of the computer.

[4] This freedom is usually limited by aesthetic codes and rules. Cf. the ongoing discussion in 18th-century Germany — within the framework of mimetic art theory — on the artistic production of marvellous or fantastic elements.

[5] We find this empirical attitude represented in Novalis' fifth Dialogue. One of the dialoguing voices (A) has a profound distrust for hypotheses and claims that "a single truly observed fact is worth more than the most brillant hypothesis". His opponent (B) attacks this "common empiricism" and takes a very strong stand in defense of hypothesis (Novalis, pp. 668—9).

[6] This can be related to what Robert Kargon calls the "judicial metaphor" in the language of the experimental philosophy in the 17th century. If the experiment is viewed metaphorically as the testimony of the witness, the scientific "trial" is seriously hampered by the witness' refusal or inability to speak.

[7] "... chercher à pénétrer [la Nature] plus profondément, à lui dérober ce qu'elle cache, à créer en quelque manière, par la différente combinaison des corps, de nouveaux phénomènes pour les étudier: enfin elle ne se borne," pas à écouter la Nature, mais elle l'interroge et la presse" (Diderot/D'Alembert p. 298).

It is worthwile mentioning that the expression "l'interroge et la presse" continues the judicial analogy, although it takes as its model a very violent judicial system that does not allow the witness to keep silent, but makes him/her speak by means of interrogation and torture.

[8] "Les spéculations les plus subtiles et les méditations les plus profondes ne sont que de vaines imaginations si elles ne sont pas fondées sur des expériences exactes" (Diderot/D'Alembert, p. 297).

[9] "Descartes, et Bacon lui-même, malgré toutes les obligations que leur a la Philosophie, lui auraient peut-être été plus utiles, s'ils eussent été plus physiciens de pratique et moins de théorie, mais le plaisir oisif de la méditation et de la *conjecture* même entraine les *grands esprits* (Diderot/D'Alembert, p. 299).

[10] "... la connaissance des faits cachés dont on s'assure en les voyant, et non le roman des faits supposés qu'on devine bien ou mal, sans les chercher ni les voir" (Diderot/ D'Alembert, p. 298).

[11] According to the distinction made by Judith Schlanger (pp. 181—202) between formal and metaphorical transpositions of scientific models into literature, this transfer works on both levels.

[12] Which is, again , a very complex case, since besides transforming the genre as such, Musil also represents experimental activity thematically in his novel *The Man Without Qualities.*

[13] "J'avertis donc qu'il est très-important de se mettre exactement à la place de la statue que nous allons observer" (Condillac, p. 222).

Mach, E.: *Knowledge and Error: Sketches on the Psychology of Enquiry* (Intro. by E. Hiebert), Reidel, Dordrecht, 1976.
Marivaux: 'La Dispute', *Théâtre complet*, Gallimard, Paris, 1949, pp. 1345—74.
Novalis (Friedrich von Hardenberg): *Schriften, vol. II: Das philosophische Werk I*, Wissenschaftliche Buchgesellschaft, Darmstadt, 1981.
Rousset, J.: *Forme et signification*, José Corti, 1962.
Schlanger, J.: *L'invention intellectuelle*, Fayard, Paris, 1983.
Searle, J.: 'The Logical Status of Fictional Discourse', *New Literary History* **6** (1975) pp. 319—32.
Schöne, A.: *Aufklärung aus dem Geiste der Experimentalphysik. Lichtenbergsche Konjunktive*, C. H. Beck, München, 1982.
Zola, E.: 'Le roman expérimental', *Oeuvres complètes*, vol. 10, Cercle du Livre Précieux, Paris, 1968, pp. 1175—1203.

Université de Montréal

science into poetry never gained institutional recognition and must be seen, from today's point of view, as a historical alternative that did not become historical reality.

[23] "— nous devons modifier la nature, sans sortir de la nature" (p. 1180)

"... instituer une expérience, pour analyser les faits et s'en rendre les maîtres!" (p. 1181)

"nous verrons qu'on peut agir sur le milieu social en agissant sur les phénomènes dont on se sera rendu maître chez l'homme" (p. 1184)

"l'homme tout puissant aura asservi la nature" (p. 1188)

"l'utilité pratique et la haute morale de nos oeuvres naturalistes: ... il n'y aura plus qu'à agir sur les individus et sur les milieux, si l'on veut arriver au meilleur état social" (p. 1188)

[24] "in die gesellschaftliche Entwicklung tief eingreifen können" (p. 297)

"jene ... auf Veränderungen ausgehende, auf Meisterung der Natur abzielende Haltung" (p. 299)

"der Zuschauer ... als der große Änderer, der in die Naturprozesse und die gesellschaftlichen Prozesse einzugreifen vermag, der die Welt nicht mehr nur hinnimmt sondern sie meistert" (p. 302)

[25] Both authors indeed make use of mechanistic metaphors in their description and analysis of society.

REFERENCES

Brecht, B.: 'Über experimentelles Theater', *Gesammelte Werke*, vol. 15, Suhrkamp, Frankfurt am Main, 1967, pp. 285—305.
Buffon (Georges-Louis Leclerc): *De l'Homme*, Maspéro, Paris, 1971.
Condillac, E. de: *Oeuvres philosophiques de Condillac*, vol. I, Presses Universitaires de France, Paris, 1947.
Diderot/D'Alembert: *L'Encyclopédie ou Dictionnaire raisonné des arts et des métiers*, vol. VI, Briasson, David l'aîné, Le Breton, Durand, Paris, 1756.
Eibl, K.: *Kritische Literaturwissenschaft*, Fink, München, 1976.
Habermas, J.: *Erkenntnis und Interesse*, 2nd edition, Suhrkamp, Frankfurt am Main, 1973.
Habermas, J.: 'La modernité: un projet inachevé', *Critique* **413** (1981) pp. 950—57.
Hartung, H.: *Experimentelle Literatur und konkrete Poesie*, Vandenhoeck und Ruprecht, Göttingen, 1975.
Kargon, R.: 'The Testimony of Nature. Boyle, Hooke and Experimental Philosophy', *Albion* **3** (1971) pp. 72—80.
Lavoisier, A.-L.: 'Discours préliminaire' to the *Traité élémentaire de chimie*. In *Pages choisies*, Editions sociales, Paris, 1974.
Mach, E.: *Erkenntnis und Irrtum. Skizzen zur Psychologie der Forschung*, first edition 1905 (rpt. Wissenschaftliche Buchgesellschaft, Darmstadt, 1976).

of knowledge, or at the very least a field of inquiry. Upon this act is necessarily imposed the establishment of a right of association. Science and literature cannot be thrown together on a *de facto* basis, as if they formed but one of a number of interesting couplings — e.g., literature and geography, science and music, etc. To the extent that it offers itself as a legitimate discipline (and this of course need not be desirable) such a discursive practice must seek instead a *de jure* foundation. Moreover, and particularly with respect to already established disciplines like science and literature, such a foundation must not take the form of a simple acceptance within the intellectual community; rather, it must be manifested as an *intrinsic moment* in the *theoretical* determination of its field of inquiry and the objects of its concepts. This my first element: the *juridical* distinction between *de facto* and *de jure* approaches to the problem of the grounding of a discursive practice.[2]

As I have suggested, the very existence of The Society for Literature and Science demands as its own founding principle the conviction of identity in language. Yet by what right is this principle asserted? Does it depend merely upon the 'facts' of language (metaphorical, metonymical, or otherwise) in the same way that a more pragmatic assertion depends upon the 'fact' that science and literature are but human activities like any other? This is not to say that such 'facts' are not important, but to question whether they can really serve to 'subvert' or 'displace' the established disciplines. This would be especially true for science, whose immanent claims to objectivity, as Foucault himself insisted in his last interview, are *in no way* impaired by their varying relations to structures of power.[3] Is it not then possible to reflect upon science and literature according to their *specific* modalities even as one acknowledges their common origin in the 'substance' that is language? Is it not possible, for example, to try to determine the *difference that is science* in or as its mode of expression, a difference which must be something more than an 'ideological' or 'mythical' moment of its own self-reflection?

The assumed validity of the above questions gives me my second element. I require access to a *differential typology* in order to specify, as far as possible, the modes of expression we call science and literature. For this task I turn to Kant. His peculiar merit with respect to these difficulties seems to me to lie in his insistence upon, first of all, a rigorous separation of validity claims according to their *de jure* grounding, and second, upon the *visibility* of these differences within the

ROBERT KOCH

HYPOTYPOSES

I. INTRODUCTION: SCIENCE AND LITERATURE AS MODES

The logic which governs the present essay takes its orientation from the specific context of its composition and presentation. This context involves an *event* and a *topic*: an academic conference held by the Society for Literature and Science, its theme "Literature and Science as Modes of Expression". At issue here are those questions surrounding what Foucault has called a discursive practice — questions of concept formation and their objects, questions of legitimacy and employment, questions of the speaking subject and its enunciations. In the particular case of this conference all such questions coalesce in a unique way; for here the event of the conference *is* its topic: what is being asked is the possibility of a new discursive practice or discursive formation 'between' those of science and literature, which necessarily implies or even requires that these disciplines be viewed as having a common ground.

Now, heretofore both science and literature have been impelled towards one another by way of various theoretical developments within each discipline. These histories are by now familiar to us. Such developments have led to the internal dislocations and conceptual re-thinkings which have prepared or opened up the space we now inhabit: this is the space of *language*,[1] which, whether through its figural, metaphorical, or rhetorical qualities, and whether through its concrete embodiment in economic, material, or social texts, in fact constructs the very worlds it purports to describe. So considered, science and literature need not beckon to one another across an unbridgeable abyss; so considered, science and literature can be thought according to their identity *in* language, their differences being, as the theme of the conference suggests with a Spinozistic resonance, merely modes of expression.

Such are the positive conditions of the context as I envision them. I wish to isolate from them two elements. The first derives from the problem of grounding. To initiate a discursive practice within the institutional framework of the university means to demarcate a domain

has nothing to do with truth. Science concerns itself with *objective validity*, a far different thing: its discovery, its justification, its *creation*. This crucial oversight explains the almost total irrelevancy of analytic philosophies of science (whether in defense or in criticism) to the development of science itself; as I intend to argue, it can explain as well the tendency for a critique such as 'science-literature' to form itself as an enclosed discursive practice.[4]

In what follows, then, I will attempt show how Kant tries to think through the unique historical phenomenon that is modern science, and to extract or invent a new image or sign that best exemplifies its reality. He gives it many names: the synthetic *a priori*, the transcendental logic, or, as I prefer, the determinant judgment. Only by fully understanding *this* reality will a discursive practice like 'science-literature' be able to create, in a *de jure* fashion, the signs of its own reality.

II. SCIENCE AS DETERMINANT JUDGMENT

A. *The Typology of Judgments*

In seeking to sketch out Kant's semiotic we must keep in mind that, strictly speaking, there can never be for him any question of the sign *as such*; and this for two essential reasons. Kant's doctrine of appearances, his phenomenology, must be understood as a description of a set of *pure relations*.[5] On the one hand, then, the sign 'as such', the sign 'in itself', is no longer meaningful. As appearance or representation the sign is a *singular* and *indivisible* affection of intuition, but given in or as a manifold. This relational character of the sign necessarily follows from Kant's description of sensibility in general: it is receptive, synthetic, it yields the experiential or empirical element in all knowledge. On the other hand this relational text of signs is never simply what Kant calls *discursive* (a technical term: it does not carry Foucault's connotations which were employed earlier). Here would be where Kant diverges from a thinker like de Saussure and indeed from the entire body of thought thereby engendered. The term 'discursive' is always used by Kant in the context of what he calls formal logic. Here the formula is: a sign refers to a sign which refers to a sign . . . etc. Kant opposes this metonymic conception by noting that the problem does not lie in the relation *between signs* but between a sign and its *mode of apprehension*. A purely discursive text of relations must presuppose an plurality of modes of expression themselves. I thus undertake what might be broadly termed an exposition of Kant's theory of signs. This will entail in the present instance a restriction to three considerations: Kant's typology of judgments, the 'pedagogic' structure which constitutes the essence of objective knowledge, and finally a typical application of this structure as found in his derivation of the concept of absolute space in the *Metaphysical Foundations of Natural Science*.

But what will be gained by such an exposition? And how will it illuminate the question of science and literature as discursive practices? More importantly, does not the couple 'science-literature' (as I shall henceforth call it) deny, and deny essentially, the claim that science can establish a *de jure* foundation? That the *de facto-de jure* distinction is tenable? Is such a denial not a manifestation of the rhetorical or linguistic component of knowledge that science has ever repressed and to which it has ever blinded itself?

As a way of responding to these questions let me anticipate somewhat the demonstration that will follow by setting forth its intellectual context. Briefly, then: for Kant the specificity that is modern science utilizes the *de jure* moment of its own self-grounding as the very *condition* by which any objects whatsoever are given; the condition, in fact, of objective validity itself. All scientific claims to validity occur within this juridical milieu. Now in itself this is not a novel assertion. Its acceptance was the primary thesis of German Idealism, its rejection or modification the common starting point for phenomenological thinkers like Husserl and Heidegger. *It stands as well as the explicit presupposition of all critique, all critical theories* (including those of literature). From Marx to Adorno to Derrida to de Man, such thought affirms in an ambiguous manner this primacy of self-grounding. Ambiguous, because on the one hand the positing of a thing as its own ground is an arbitrary and exclusionary act, a violent act; but on the other hand a necessary one, for it is the essence of consciousness as such, without which no assertions, discursive practices, or knowledge is possible. 'Science-literature' stands in this line of thinking: it affirms the necessity of self-positing, not in order to attain to the *de jure* truth of its concepts but rather in order to reveal *itself* as yet another fiction, intrinsically incapable of performing or carrying out the very task it has set for itself (or rather it can *only* perform it, 'play' or 'stage' it).

But all such 'philosophies of science' (for critique as such is necessarily a philosophy of science) miss an essential Kantian insight: *Science*

relation with intuition; (ii) he thereby *changes the concept of judgment itself*. We must recall that a thesis which asserts an infinite semiosis brings with it a certain view of judgment, in which attributes are predicated of a substantial and identical subject. This is the logical judgment. But for Kant, as we have seen, a sign refers essentially not to another sign situated on a homogeneous plane but to another plane altogether, to intuition. He does not invoke judgment as a means of arresting the 'drift of signifiers' by appealing to a 'transcendental signified'. As he is at pains to point out — a fact infrequently remarked — neither his transcendental ego nor his thing-in-itself are or can be *constitutive* or *substantial* entities. Rather, the possibility of either depends upon and is established through the *de jure* claim to validity as *evidenced in the structure of the judgment*. The transcendental ego and the thing-in-itself are but end-points, ideal extrapolations from a primary and mobile set of relations which is the judgment. I would therefore argue that one need not *necessarily* detect in Kant's explication of the judgment a subjectivism or a rigid subject-object epistemological structure. There is rather a set of relations: the set of signs given in intuition on one plane and a set of judgments which intersects them on another plane.[7]

B. *Examples, Schemata, Symbols*

Thus far I have tried to present the positive conditions whereby one might grasp the general function of signs within Kant's critical philosophy. The relation between signs and judgments becomes altered somewhat by the time he writes his *Critique of Judgment*. It is now a question of the relation between *universals* (concepts, principles, rules, ideas) and *particulars* (cases, empirical intuitions, and their employment). Let me turn to this text as one of the few places in his entire work where he explicitly discusses the problem of signs. This is found in Section 59, entitled "Beauty as a Symbol of Morality". The context here is Kant's distinctions between, on the one hand, *schemata*, which immediately demonstrate the validity of pure concepts of the understanding; and *symbols*, whose agreement with pure concepts of reason is brought about by analogy. Both schemata and symbols Kant calls *hypotyposes*, which he defines in the singular as a presentation, a "rendering in terms of sense" [*Versinnlichung*] (1911, A252/B255–6).

underlying homogeneity among what Peirce has called the sign, the referent, and the interpretant (thus the subject is but another sign). Kant insists to the contrary that the sign involves and expresses an *ostensive* relation, a relation to an 'other': not this or that other, but to the other of thought, i.e., to intuition, our mode of being-affected. *For Kant this mode is never a sign*: it is a pure 'relation-to', and lies on a different plane than the signs of discursive thought; it is the very possibility of all signs.[6]

The second essential reason follows from the above. If we again refer to Peirce (whose own typology of signs bears a certain resemblance to Kant's), except for the purpose of analysis one can no longer distinguish between the sign in its material quality, the sign in relation to the real, and the sign in relation to its employment (see Pharies). For Kant these aspects are inseparable. The sign is always given only as embedded within or as a mode of demonstration or a demand for legitimation — as a *judgment*.

That Kant laid the function of judgment at the heart of his critical system is a common observation (See Buchdahl, ch. 8, and Nachelmans). It is a move that has often been deplored, since it seems to restrict human knowledge of the world in the direction of Newtonian physics. Is it really the case that all human knowledge and the signs employed therein must be modelled upon the structure of the judgment? But it may be that such objections in turn restrict too narrowly the meaning of judgment for Kant. We can perhaps glimpse this wider understanding in the sheer multiplicity of those he enumerates: he distinguishes between judgments of perception and experience; between logical, practical and cognitive judgments; between analytic and synthetic, *a priori* and *a posteriori* judgments; between aesthetic and teleological judgments; and between determinant and reflective judgments. But throughout all of these distinctions Kant labours to uncover the answer to a single question: By what right and through which means can such and such a sign be referred to the intuition of an object? In other words, what happens in a given judgment? What forces are set into motion? What *must* happen if a claim is to have objective validity? As we know, the answers Kant came up with have proven historically determinative, especially with respect to knowledge claims found in aesthetic judgments.

The point I wish to stress here is that by laying judgment at the base of signs Kant does two things: (i) he brings judgment into an essential

turn momentarily. I shall reserve until the conclusion further comments about symbols and the reflective judgment. (iii) Why 'hypotyposis'? The very fact that Kant chooses this term is curious. In antiquity Quintilian employed it in the formal sense as a rhetorical figure meaning 'vivid impression', a lively depiction of a scene or event. It also had the more general connotation of 'plan' or 'sketch', as in Sextus Empiricus' *Hypotyposes (Outlines of Pyrrhonism)*.[9]

At this conjunction of rhetoric and philosophy we encounter the 'science-literature' line of critique. For someone like Eugenio Donato the function of hypotyposis is to "produce a visual pictorial image"; its significance is that it "threatens the capacity of a linguistic representational mode to dissolve its objects into abstract idealities by transforming the letters into belated visual entities". Here a certain 'property' of language (its visual, figural character; it could just as well be its metaphoric or metonymic quality) resists a certain conceptual appropriation. And of course Donato infers from the resistance of such figural play that the distinction between literary and philosophical (and presumably scientific) discourse is "untenable". A distinction which is no distinction: upon this argument is founded the very possibility of a new discursive practice or a new kind of knowledge that is neither 'scientific' nor 'literary' or is both at once.[10]

There are many difficulties here, not the least of which concerns the status of this resistance, this so-called 'subversive' nature of language (If Donato does not point it out for us, do figures still resist conceptual appropriation? If so, where? And for whom? Themselves? In what space do subversions make a difference? Merely within critical discourse?). In the section which follows the possibility will be raised that for Kant at least science is a discursive practice whose objects and concepts are grounded in *just this displacement of the rhetorical or discursive* elements of language. The schemata which belong to the determinant judgment have no essence or intrinsic characteristics: they are a *function* of a prior and enabling *de jure* act.

C. *The Transcendental Pedagogy*

For Kant nature is a "nexus of rules" (1974, p. 13). As he notes in the Preface to the Second Edition of the *Critique of Pure Reason*, modern science begins with the discovery that this nexus must in fact be subjected to the principles of judgment (1965 B xiii). Given the fact

There are yet other distinctions. If the concepts are empirical as opposed to pure, then the corresponding intuition or sign is termed an *example*. Yet the example too is a hypotyposis; it possesses an intrinsic connection to an intuition. In contrast, "marks" [*Charakterismen*] are but designations [*Bezeichnungen*] or mere expressions [*Ausdrücke*] of concepts, such as we find in algebraic or mimetic signs. Kant stresses in a note to the same passage that both modes of knowledge are intuitive and must be opposed to a discursive mode.

Some comments on this exposition. We see displayed Kant's penchant for demarcation and for the establishment of precise domains of authority. Once again, however, it is possible to see in this something more than an activity of exclusion and repression: the workings of a differential 'motor', if I may be excused the phrase. For our present purposes let me make three observations. (i) The distinction between schemata and symbols, especially with respect to their methodological employment, is hardly comprehensible unless one is aware of the specific roles Kant assigns to the understanding and to reason. The understanding is the faculty of rules and concepts; its proper role is to enact what Kant calls a *legislative authority* over the signs that come within its purview. If these signs are merely discursive we have marks and a system of formal logic. If they are given in intuition we have schemata and the possibility of objective knowledge. But if the concepts of rules of the understanding attempt to go beyond these domains, if they attempt to legislate symbols rather than marks or schemata, there arise Kant's famous antinomies. Only reason, as that faculty concerned with the unconditioned element in knowledge, is permitted to employ symbols. (ii) From the above it is clear that scientific or objective knowledge in the strictest sense (what Kant calls the synthetic *a priori* knowledge of objects) is limited to those hypotyposes he terms schemata and the form of judgment which rules them. As we shall see, for Kant mathematics encompasses such hypotyposes. But it is necessary to recognise just how large the structure of this form of judgment looms in the later Kant. Indeed, by the time of the third *Critique all* forms of judgment had been reduced to two: the determinant judgment, which employs schemata and subsumes *a priori* the cases which come before it; and the reflective judgment, which employs symbols in its search for universal principles. *The structure of scientific knowledge is thus the structure of the determinant judgment.*[8] It is to this structure that I shall

tion of a pre-predicative similarity or affinity. *But this is not the recognition of a discursive affinity*, i.e., a recognition of an *identity* subsisting between the particular cases themselves. It is rather the recognition of an *ostensive* affinity between the synthetic 'relation-to' element of intuition and the rule of judgment. The *de jure* positing of the rule is now the real ground for the very appearance of the sign itself. *The sign comes into existence only as the object of a rule.*

Hence the importance for Kant of mathematics. However dubious Kant's insistence that mathematics *constructs* its concepts will appear to later set theoreticians,[11] for him its significance lies rather in its peculiarity as a mode of cognition. Mathematical signs are a curious form of hypotyposis. On the one hand, as we have observed, the externality of algebraic signs to intuition makes them interchangeable and thus mere marks. On the other hand, geometrical objects seem to be able to represent their concept or universal rule empirically "without impairment" as he says; here the universal *is* the particular (Kant, 1965, A713—38/B741—66). I cannot explore the intricacies of this problem within the scope of this paper except to note that the ambiguous status of the particular sign is brought about, once again, because it is given only as a *function* of the synthetic unity of the act of understanding which generates it.[12] The significance of mathematics for Kant seems to lie rather in the fact that it is that form of knowledge in which *all judgments are determinant*. It no longer has to distinguish *whether* a universal rule applies to a particular case since it *a priori* constructs the cases themselves.

A point of discussion. Kant asserts that examples are capable of verifying only empirical concepts and can yield only 'formulas' or maxims and not objective principles. Yet, it will be objected, how is it that Kant's own text is *forced to employ* the example of a teacher? What then becomes of its own claims? Does not the very 'textuality' of the text (in this case, the use of examples to deny their validity) betray its own claims to truth?[13]

There is no denying it: Kant's text presents us with a performative contradiction. But this is merely an *analytic* question: *the question here is not that of truth*. What is at stake is rather the *synthetic* possibility of objects of experience. The next section will try to explicate the differences between these approaches. One can see, however, that the concept of 'textuality' remains essentially within a discursive milieu, the milieu of formal logic. The value of contradiction is at once upheld as

that the rules or principles of the understanding are wholly heterogenous to those of nature, or to the intuition which is affected by them, how can such a subsumption occur?

For Kant it depends upon the possibility of what he calls a *transcendental* logic, which he opposes to mere formal or general logic. This latter is characterized by what I have earlier noted: it comprises a purely discursive relational text of signs or representations. The judgments proper to formal logic, then, have no intrinsic application whatsoever to objects of experience. *In this light* their employment appears arbitrary and subjective, "a peculiar talent that can be practiced only, and cannot be taught". Judgments belonging to formal logic *cannot be taught*; conversely, in this sphere the understanding (which again for Kant is nothing but the faculty of rules) *cannot teach*: "General logic contains, and can contain, no rules for judgment" (1965, A132/B171). A strange pedagogical scene: an understanding which cannot teach, judgments which cannot be learned. And it is because the rules of the understanding here have no immanent connection with experience that their application should seem but an unprincipled and wayward 'art', a mere playing with marks.

But it is here where I believe Kant's penetration into the logic of modern science can be located. At this point he gives an example of teachers unable truly to teach and details the reason: "[They] may comprehend the universal *in abstracto*, and yet not be able to distinguish whether a case *in concreto* comes under it" (1965, A134/B173).

At stake is indeed what may be called a *rhetorics* of conceptual deployment. By this I mean, in a sense similar to Nietzsche's, the relation that concepts or rules hold with respect to their objects, that manner in which concepts *stand over* them (see Blair). Now at least since Spinoza it had been recognized that the objective validity of an idea cannot be grounded merely in its correspondence to its objects, in its *truth*, for then its validity would possess only a *de facto* status. Rather, as Kant now conceives it, validity must be grounded in the mode of representation itself, or more specifically in the legislative authority of the application of an *a priori* rule, as the very condition whereby the object in fact becomes possible. *The rule and the ruled are given together.*

How is this accomplished? *Prior* to the rule being applied one must first have ascertained *that* the rule being applied is in fact applicable; *prior* to all application whatsoever comes the act of *seeing*, the recogni-

How is this? A metaphysics of corporeal nature implies the *construction of the real* in appearance: sensation or intensive magnitude, that which is properly empirical in all signs. Now in the *Critique of Pure Reason* this element of the real differentiated even on a methodological level objects of pure mathematics and objects of experience. To be sure mathematics is paradigmatic of knowledge as such, it enacts the determinant judgment. Thus Kant will state that "the formative synthesis through which we construct a triangle in imagination is precisely the same as that which we exercise in the apprehension of an appearance" (1965, A224/B271). And yet of course the *Methodenlehre* of that text concerns itself with distinguishing between philosophy (science) and mathematics in regards to their employment and the objects of their knowledge. Human knowledge of the world can never be truly mathematical for it must wait for appearances or signs to be given in intuition; the real in sensation cannot be constructed, merely anticipated.

But this is why the concept of *cause-and-effect* becomes so crucial for Kant, and why it stands at the very heart of all objective knowledge. Although the real cannot be constructed, and thus made into a mathematical object, the *structure* of the determinant judgment given in mathematics is carried over, via the schema and as a determination of time, into the apprehension of appearances in intuition. Kant: A follows B "*necessarily and in accordance with a universal rule*"; its "dignity" lies in the fact that "the effect not only succeeds upon the cause, but that it is posited *through* it and arises *out of it*" (1965, A91/B124; Kant's emphases).

The concept of cause-and-effect is thus the form that the mathematical or determinant judgment assumes in its apprehension of the real in appearances. As Kant explains it is a *dynamical* concept and not a mathematical one. If we return now to the *Metaphysical Foundations* we will see that what begins to occur is that the dynamical concept conflicts with the mathematical. Indeed, I propose that, just as mathematics requires and thus posits the construction of objects in intuition, in a heterogenous plane, as a means of breaking with the discursive and tautological plane of formal logic, so does the "metaphysico-dynamical" mode of explication break with the similarly uniform plane of pure mathematical objects. What is essential is thus not the planes themselves (for they each have their specific modes and objects) but rather the construction of another plane, another dimension, another virtuality. *For Kant this movement is nothing else than the movement of*

the standard of measure and derided as an impossible ideal. It is the milieu of tautology. We can thus glimpse the significance of Kant's understanding of science in a more comprehensive way. The displacement of 'textuality' in general (of the figure, image, metaphor, of their so-called "materiality') indicates both conceptually and *historically* what may be termed the *destruction of tautology*, whether it is a feature of Scholastic logic or the 'quasi-analytic' character of scientific theories themselves (Kuhn, note p. 304). Such destruction is achieved precisely through the extraction or realization of this pure synthetic structure that Kant calls the determinant judgment, which has many analogues in other thinkers: Kuhn's notion of 'symbolic generalisation', Leibniz's 'enchantillons', Gerald Holton's 'propositional space', Gilles Deleuze's 'any-space-whatever', or Bachelard's 'super-object'.[14] What is common throughout is the insight that scientific objectivity is tied to the freeing up of a pure virtuality, a pure synthetic 'substance' which breaks essentially with the analytic conundrum.

D. *The Phenomenological Sign and Absolute Space*

I wish to show now how the structure of the determinant judgment works in Kant's *Metaphysical Foundations of Natural Science*. As usual he sets forth his project by attempting to ascertain the legitimate objects of his inquiry; and once again at issue is the relationship between a mere art (even a rational one, like the chemistry of his time) and what he terms science proper. But this relationship has another dimension: the problem of the *general* and the *special*. The structure of the determinant judgment as isolated in the first *Critique* always carried with it an essential restriction: it made possible only the experience of nature *in general* and thus remained "undetermined regarding the nature of this or that thing of the sense-world" (1970, p. 469). The possibility of a *special* metaphysical natural science would seem to be possible only by searching the 'range of cognition' of which reason is *a priori* capable with respect to given *empirical* concepts. In other words, a special metaphysical science would have to deal with examples. But Kant answers with his well-known assertion that "in every special doctrine of nature only so much science proper can be found as there is mathematics in it" (1970, p. 470). Kant is insisting upon a metaphysics of *corporeal* nature (not nature in general) which, strictly speaking, is impossible given the parameters of his system.

absolute impenetrability or something like a repulsive force. For Kant the former was an 'occult' quality; the latter, although it too had to be admitted as fundamental and thus not capable of further explanation, "nevertheless yields the concept of an active cause", the effect of which (resistance in filled space) "can be estimated according to the degrees of this effect" (1970, p. 502). Kant treats mechanics in a similar manner. When he attempts to construct Newton's first and third laws of motion he encounters a difficulty since matter so conceived seems to have no other quantity than that found in the sheer multiplicity of parts external to one another. Thus it makes no difference whether the *degree* of moving force (i.e., the real in perception) is determined by a quantity of motion or a quantity of matter. Again we encounter the problem of 'discursive indeterminacy', so to speak. Kant notes however that this "reciprocal derivation of two identical concepts from one another" results from the mere *analysis* of their content and neglects the *application of the concept to appearances* (1970, p. 539). If this is done, the *synthetic* concept of the conservation of matter yields the requisite determinacy.

The chapter on phenomenology stands as a succinct summary of the above points and will lead into some concluding remarks. Kant wants to justify positing something like absolute space at the foundation of his special doctrine of determinate nature. He grants that all appearances and all motions as given in appearances are merely relative. But, he says, in order to *think* motion *for the sake of a possible experience* one must be able to indicate the conditions under which matter is in any way determined; one must be able to ascribe the predicate 'motion' to an intuition. He then identifies the crucial element: "Here the question is not of the transformation of illusion [*Schein*] into truth, but of appearance [*Erscheinung*] into experience" (1970, p. 555).

Was Kant wrong in positing an absolute space? What does 'wrong' mean here? The objective validity of a scientific concept is not a matter of its truth: it is whether or not it can legislate according to the structure of the determinant judgment the appearances in intuition. It is a matter of transforming marks into schemata, of inventing new conceptual planes whereby analytic tendencies are thwarted. When David Bohm invents the notion of an 'implicate' or 'generative' order, is he not doing this very thing? Is he abandoning science for metaphysics or is he in fact continuing along the line that belongs to science and to science alone?

science itself. It is *itself* in its movement its own *de jure* validity. Let us follow its line as Kant traces it.

For natural science matter is the real in all appearances; but what is sensed is not matter as such, since all appearances are purely relational, but matter in motion. Natural science is thus the doctrine of the motion of matter. Kant brings this motion into the structure of the determinant judgment (here, the pure concepts of the understanding as given in the table of categories) yielding the following divisions: phoronomy, or kinematics, motion as pure quantum; dynamics, or motion as quality; mechanics, or motions in relation; and phenomenology, or motion considered with reference to its modality, i.e., as an appearance of our external sense.

Phoronomy has as its object the construction of motion as a quantity; it therefore admits of a purely geometrical explication. It represents the motions of a point according to their direction and velocity, but these considered not as effects of physical causes but merely from or according to "rules of congruity". This follows from the fact that in itself geometrical space is uniform and that the composition of motion through identical quantities can never regard any one quantity as determinate. *Hence all motion is relative.* All motion or rest, all bodies in motion as well as the space they fill, are here all given as signs or appearances; one can just as easily view the bodies as resting and set space into motion. Now the difficulty here for Kant does not lie in any recoil at the prospect of relativity as such. He readily grants that the idea of an absolute motion is absurd. But to leave things at this level is to fail in the proper task of science itself. In a curious way it is as though phoronomy remains *too logical*, too dependent upon identity. The indeterminacy of marks must be transformed into determinate schemata or, because we are here dealing with empirical concepts, into 'schemata-examples'.

As Kant moves from phoronomy to dynamics and then to mechanics, then, his aim in each case can be seen as a *further determination* of an object according to the virtuality proper to science. The conflict which arises in the chapter on dynamics between mathematico-mechanical and metaphysico-dynamical modes of explication, which of course has its source historically in the debate over Newton's postulation of universal gravitation, must be seen in this light. For Kant only dynamical grounds "admit the hope of determinate laws" (1970, p. 534). In specific terms the debate was couched as an appeal to either

of its virtual power? What tautologies does it destroy, what kind of a plane does it construct? Perhaps that point of intersection, which I have called 'science-literature', is nothing other than the reflective judgment, and the symbol its proper hypotyposis.

NOTES

[1] See the excellent introduction to Jordanova.
[2] On the question of discursive practices in general, see Foucault (1972); on the necessity for an intrinsic theoretical moment, see J. Habermas, 'The Idea of the University', *New German Critique* **41** (1987) pp. 3—22; on the *de jure/de facto* distinction, especially with regard to Kant, see Smyth, pp. 98—100.
[3] R. Fornet-Betancourt, H. Becker, A. Gomez-muller, 'The Ethics of Care for the Self as a Practice of Freedom: An Interview with Michel Foucault on January 20, 1984', trans. by J. Gauthier, *Philosophy and Social Criticism* **12** (1987) p. 127. For a different perspective on Foucault, and a more common one, see D. Thompson, 'Epistemology and Academic Freedom', in *Descriptions* (ed. by D. Ihde and H. Silverman), SUNY Press, New York, pp. 286—95. But see also Foucault (1980), pp. 109—10, with respect to what he calls the 'thresholds' of possible explanations according to their 'epistemological profiles'.
[4] Thus Hans-Georg Gadamer will make a useful and typical insight in his essay "The Philosophical Foundations of the Twentieth Century": one can view most intellectual schools and developments in the century as critiques of the natural sciences and their methodology. One must certainly account for this: one must also account for their *superfluity*, even and especially in the case of a 'favorable' school like positivism.
[5] Martin (1955) stresses Kant's affinity here to Leibniz.
[6] For a variant of this argument with regard to the question of the homogenous, see C. Levin, 'La Greffe de Zele: Derrida and the Cupidity of the Text', in Fekete (1984), pp. 201—27.
[7] Kant discusses the transcendental ego and thing-in-itself at (1965), A250—53; (1978), note p. 14; (1970), pp. 542—3. See also a penetrating comment by Fernandes, p. 141. Heidegger's interpretations of Kant have particularly emphasized the heterogenous factor of intuition (1962, 1965).
[8] A point to be noted in passing. The structure of the determinant judgment holds as well for the application of the *ideas of reason*, although part of their function now passes to the reflective judgment. Both the syllogism and the practical judgment are essentially determinant. For an opposing view cf. Nancy, who sees moral judgments as breaking with the apophantic mode. George Schrader gives a fine analysis of the role of the reflective judgment in his classic article 'The Status of Teleological Judgment in the Critical Philosophy'.
[9] See Quintilian, *Institutio Oratoria*, VIII, iii, 61ff.; IX, ii, 40—44; and Cicero, *De Oratore*, III, liii, 202. See also in this respect Bruns and Leff.
[10] De Man earlier purused a similar line of critique.
[11] Martin (1985) provides an excellent survey and discussion of the problem of construction in Kant.

III. CONCLUSION: DYNAMICS AND THE SYMBOL

It is beyond the scope of this paper to explore in greater detail the implications of the questions raised above. My purpose has been simply to open up a certain space for reflection. And it is to the problem of reflection — to reflective judgment — that I now turn by way of a conclusion.

Let me stress once more Kant's insight into modern science: it does not concern itself with the question of truth. No adequate understanding *or* delimiting of its force can be obtained unless this is recognised. The strategies which have thus far dominated the line of critique I call 'science-literature' seem to me to be fundamentally intertwined with this question, which, as I have suggested, they confuse with that of objective validity. To suggest that claims to objective validity as given in science are undermined by exposing the ineluctable historicity of their context, or the metaphoricity of the language in which they are articulated, or the genealogical conditions of power in which they are entangled, is to miss the point. Such a strategy has *never touched* and *will not touch* that complex body of discursive practices we call modern science; at most it will create for itself a separate academic discipline alongside others, committing itself to the elaboration of its own conceptual objects and disciplinary aims. But — as indicated perhaps by the conference itself — it will have nothing to say to science.

From my viewpoint the point of intersection between the disciplines of science and literature must be located elsewhere than in or as an analytic critique of presuppositions. Perhaps Kant's own work provides us with a possible line of trajectory. We have thus far had little to say about the reflective judgment or the symbol which is its proper sign. But one can glimpse something of the field opened up by these notions, of which I give a brief itinerary: the problem of *intensities*, of pleasure and pain, beauty and the sublime, of taste and communicability; the problem of *experience*, the experience of the real in or as the formative power of nature, or nature as a system, the experience of experientiality itself; the problem of *reflection*, of thinking as a social habit embedded within a common space, a common sense. All these problems are united in what Kant calls the symbol and the operation of the reflective judgment. Its mode of operation, as he says "has been but little analysed, worthy as it is of deeper study" (1911, see 59). What is its structure? What is its *de jure* claim? What are the features and effects

Heidegger, M.: *What is a Thing?* (trans. by W. Barton and V. Deutsch), Regnery/Gateway, South Bend, 1967.
Holton, G.: *Thematic Origins of Scientific Thought: Kepler to Einstein*, Harvard Univ. Press, Cambridge, Mass. and London, 1973.
Jordanova, L.: *Languages of Nature*, Free Association Books, London, 1986.
Kant, I.: *Anthropology from a Pragmatic Point of View* (trans. by V. Dowdell), Southern Illinois Univ. Press, Carbondale, 1978.
Kant, I.: *Critique of Judgment* (trans. by J. Meredith), Clarendon Press, Oxford, 1911.
Kant, I.: *Critique of Pure Reason* (trans. by N. Smith), St. Martin's, New York, 1965.
Kant, I.: *Logic* (trans. by R. Hartman and W. Schwarz), Bobbs-Merrill, Indianapolis, 1974.
Kant, I.: *Metaphysical Foundations of Natural Science* (trans. by J. Ellington), Bobbs-Merrill, Indianapolis, 1970.
Kuhn, T.: *The Essential Tension: Selected Studies in Scientific Tradition and Change*, Univ. of Chicago Press, Chicago, 1977.
Leff, M.: 'The Topics of Argumentative Invention in Latin Rhetorical Theory from Cicero to Boethius', *Rhetorica* **1** (1983) pp. 23—44.
Martin, G.: *Arithmetic and Combinatorics: Kant and His Contemporaries* (trans. by J. Wubnig), Southern Illinois Univ. Press, Carbondale and Edwardsville, 1985.
Martin, G.: *Kant's Metaphysics and Theory of Science* (trans. by P. Lucas), Manchester Univ. Press, Manchester, 1955.
Nachelmans, G.: *Judgment and Proposition from Descartes to Kant*, North-Holland, Amsterdam, 1983.
Nancy, J.-L.: *L'Impératif Categorique*, Flammarion, Paris, 1983.
Pharies, D.: *Charles Peirce and the Linguistic Sign*, Philadelphia, John Benjamins, 1985.
Schrader, G.: 'The Status of Teleological Judgment in the Critical Philosophy', *Kant-Studien* **45/46** (1953/1954/1955) 204—35.
Smyth, R.: *Forms of Intuition: An Historical Introduction to the Transcendental Aesthetic*, Martinus Nijhoff, The Hague and Boston, 1978.

York University

[12] The question of the empirical status of geometrical figures is linked by Kant to the problem of objective finality. See (1970), sec. 62, 'Analytic of Teleological Judgment'.
[13] This tends to be the strategy employed by Derrida when dealing with the relationship of a philosophical concept to its empirical presentation.
[14] The references are: Kuhn, pp. 293—319; Cassirer, p. 402; Holton, pp. 328—38; Deleuze, p. 4; Bachelard, p. 119. Husserl's concept of mathematization as an ideal limit-horizon is also relevant here.

REFERENCES

Note: The following references list the primary texts consulted in the writing of this essay.

Bachelard, G.: *The Philosophy of No: A Philosophy of the New Scientific Method* (trans. by G. Waterston), Orion Press, New York, 1968.
Blair, C.: 'Nietzsche's Lecture Notes on Rhetoric', *Philosophy and Rhetoric* **16** (1983) pp. 96—129.
Bohm, D. and Peat, D.: *Science, Order, and Creativity*, Bantam, New York, 1987.
Bruns, G.: *Inventions: Writing, Textuality, and Understanding in Literary History*, Yale Univ. Press, New Haven and London, 1982.
Buchdahl, G.: *Metaphysics and the Philosophy of Science*, Blackwell, Oxford, 1969.
Butts, R., ed.: *Kant's Philosophy of Physical Science*, Reidel, Dordrecht, 1986.
Cassirer, E.: *Philosophy of Symbolic Forms*, vol. III (trans. by R. Mannheim), Yale Univ. Press, New Haven and London, 1957.
Deleuze, G.: *Cinema 1: The Movement-Image* (trans. by H. Tomlinson and B. Habberjam), Univ. of Minnesota Press, Minneapolis, 1986.
de Man, P.: 'The Epistemology of Metaphor', *Critical Inquiry* **5** (1978) pp. 13—30.
Derrida, J.: *Dissemination* (trans. by B. Johnson), Univ. of Chicago Press, Chicago, 1981.
Derrida, J.: 'Economimesis', *Diacritics* **11** (1981) pp. 3—25.
Derrida, J.: *The Truth in Painting* (trans. by G. Bennington and I. McLeod), Univ. of Chicago Press, Chicago, 1987.
Donato, E.: 'The Writ of The Eye: Notes on the Relationship of Language to Vision in Critical Theory', *Modern Language Notes* **99** (1984) pp. 959—78.
Fekete, J., ed.: *The Structural Allegory: Reconstructive Encounters with the New French Thought*, Minneapolis, Univ. of Minnesota Press, 1984.
Fernandes, S.: *Foundations of Objective Knowledge*, Reidel, Dordrecht, 1985.
Foucault, M.: *The Order of Things: An Archaeology of the Human Sciences*, Tavistock, London, 1970.
Foucault, M.: *Archaeology of Knowledge*, Tavistock, London, 1972.
Foucault, M.: *Power/Knowledge: Selected Interviews and Other Writings 1972—77* (ed. C. Gordon), Pantheon, New York, 1980.
Gadamer, H. G.: 'The Philosophical Foundations of the Twentieth Century', in *Philosophical Hermeneutics* (trans. by D. Linge), Univ. of California Press, Berkeley, 1976, pp. 107—29.
Heidegger, M.: *Kant and the Problem of Metaphysics* (trans. by J. Churchill), Indiana Univ. Press, Bloomington, 1962.

alchemical treatises such a narrative terminus receives little detailed attention. In effect, the ultimate closure that alchemy would have us imagine is absent. Instead of a single final event that receives sole attention, one discovers a process involving many stages each with a particular closure. The alchemical process may be compared to reading a novel with a complex plot in many chapters. George Ripley, for example, divides his *Compound of Alchymie* into twelve steps: calcination, solution, separation, conjunction, putrefaction, congelation, cibation, sublimation, fermentation, exaltation, multiplication, and projection.[3] As in reading a novel, each individual section becomes plotted within a larger chain. Alchemical emblem books by figures such as Michael Maier or Basilius Valentinus offer an even more vivid example of the sequential presentation of the process.[4] The image of the chain of being offers a useful way of thinking about this process, especially if we think of the formation of each link as one level of narrative and the image of the completed chain as the overarching or metanarrative that informs the entire sequence. (Since metanarrative has come to indicate a shared perception of the general process, I find its use appropriate here.[5] In its simplest form the alchemical metanarrative is a story of perfection.) We must be careful not to give sole importance to the metanarrative for the simple reason that it is often the local narrative rather than the metanarrative that receives the most detailed attention. Above all we need to acknowledge that alchemy proceeds by employing a metanarrative and a series of local narratives.

We may analyze the form of alchemical narratives further by comparing them to the common rhetorical figure known as the *exemplum* or enthymene. Although we have come to expect that examples direct an audience to a simple closure, they provoke another response as well. Remarks in the *Rhetorica ad Herennium* are typical: "*Exempla* are not distinguished for their ability to give proof or witness to particular causes, but for their ability to expound these causes" (Battaglia, p. 455 [my translation]). Often we mistakenly assume that an exemplum would simplify discourse when it actually works towards its complication. Such a function reminds us how misleading it is to reduce fables in sixteenth-century European literature to the conclusions attributed to them in mythographic handbooks. Conclusions are not end points alone but points of departure. A similar point can be made in regard to alchemical narrative. While the closure of an experiment certainly bears significance, it is only a stage in an evolving process. Rather than

KENNETH J. KNOESPEL

THE MYTHOLOGICAL TRANSFORMATIONS OF RENAISSANCE SCIENCE: PHYSICAL ALLEGORY AND THE CRISIS OF ALCHEMICAL NARRATIVE

The movement from alchemy to chemistry may be viewed as a paradigm shift, but such characterization simplifies the extraordinary complexity of the transition and blurs the ways alchemy's preoccupation with narrative continues to influence the new discipline. A register of such change appears in the *Philosophical Transactions of the Royal Society*, where one moves from the marvelous stories of monster births in the earliest volumes to the requests for astronomical data and soundings from the south seas in later ones.[1] Transitions, displayed in passing in the *Transactions*, stand out in alchemy where multiple narrative strategies on the local as well as metalevel come into play.

It is a common assumption that the shift from alchemy to chemistry amounts to a transition from a narrative to an atomistic semiotic code. Stripped of its scientific pretension, alchemy appears, in the words of a recent scholar, as "poetic clothing for unconscious psychic processes" (Dobbs, p. 32). But we can learn more from alchemy by thinking beyond its naive sentiment or quaint visual images. Even after the seminal work of scholars such as Allen G. Debus, alchemy remains enclosed in an antiquarian reserve that may be drawn on to justify the story of a more rational inquiry.[2] Besides commenting on the local and metanarratives that comprise alchemical texts and emphasizing how alchemists sought to bring stability to their narratives, I want to suggest that the thematic interest in purification so obvious in alchemical narratives does not disappear but becomes condensed in the metaphors of purification found in eighteenth-century chemistry. My remarks conclude with several questions concerning the ways language and narrative intervene in the practice of modern as well as Renaissance science.

I

The closure symbolized by the philosopher's stone or associated with the ultimate transmutation of base metal into gold represents the most common narrative component within alchemy. Yet when one looks at

> Thy Substance thus together proportionate,
> Put in a Bedd of Glasse with a bottome large and round,
> There in due tyme to dye, and be regenerate
> Into a new Nature, three Natures into one bound,
> Then be thou glad that ever thou it found. (p. 321)

The balance between abstract invocation of theory and practical description is hardly unusual within alchemical manuals. It is especially in the sections dealing with experimentation that we find the greatest variation. Alchemical narratives vary because they permit the continuous incorporation of new narratives of experimentation. While it is possible to dismiss the proliferation of these narratives as incoherent from a modern standpoint, they also represent an interest in observing and reporting subtle changes. At the same time that alchemy's metanarrative is abstract, the localized narrative describing activities moment by moment can convey much detail. Todorov's observation that medieval romance juxtaposes two types of episodes — adventures and the description of their significance — may be applied to alchemical narrative as well.[6] The plot of the romance adventure corresponds to alchemical emplotment of experimentation, while the description of significance pertains to the alchemical metanarrative. In its distinction between theory and practice, *Bloomefield's Blossoms* exemplifies such double narrative.

We have such an enormous number of varying alchemical treatises because they are notes of individual operations and because there is no master manual that would accommodate the practical work. Moreover, the alchemical metanarrative embracing purification remains too general to guide local narratives. In contrast to mythological manuals such as Conti and Ripa, which demonstrate how carefully moral fables were controlled, alchemical treatises embody a kind of hermeneutic instability that itself promotes further experimentation.[7] Each effort bears an expectancy that *it* will bring order to a chaotic process. Even though stabilizing correspondences may be discovered with the alchemical metanarrative, their psychological nature causes them to have little bearing on the physical world. Although assimilation of the master narrative permits one to construe its local manifestations, the master narrative, because of its own abstraction, hinders the formulation of narratives of discovery on the local level.

approaching the multitude of variations in alchemical narratives as signs of disorder, we should think of them as evidence of the serious exploration of causes. At the end of *The Compound of Alchymie* (1471), George Ripley urges the reader to "stody tyll thou understond eche Chapter by and by" (p. 188). *Norton's Ordinall* closes with a testimony to the process: "If ye compleate their Sentences all,/Not by Opinion, but after this *Ordinall*" (Ashmole, p. 105). In effect, the anticipated result or expectation accompanying alchemical narrative becomes a departure point for a more significant process.

By concentrating on the process rather than a psycho-physical telos, we discover multiple components coupled to give an experimental formulation. The narrative structure of alchemy lends itself to elaboration: the form combines what we may think of as rules and their application in contrast to making them into wholly separate operations. The author of *Bloomefield's Blossoms* (ca. 1540) distinguishes between the metanarrative of evolving perfection and laboratory work by dividing the work into *Theorica* and *Practice* (Ashmole, pp. 315 and 319). The former reminds the reader of the mystery governing the experiment and is not unlike an invocation or blessing before a specific act.

> Devotely therefore unto thee O Lord I call,
> Send me thy Grace to make explication
> Of Chaos: For thou art opener of seacrets all:
> Which ever art ready to heare the Suplicacion
> Of thy meeke Servants, which with hearty humiliacion
> To thee do I apply: send me now thy grace
> Of thy Secrets, to write in due order tyme and place. (p. 316)

In the section on *Practica* we find instructions pertaining to our moral behavior: "Be you Holy therefore, Sober, Honest, and Meeke;/Love God and your Neighbour, to the Poore bee not unkind" (p. 319), as well as practical work:

> Then after that they be one Body made,
> With the sharpe teeth of a Dragon finely,
> Bring them to Dust, the next must be had,
> The true proporcion of that Dust truly,
> In a true Ballance weighing them equally;
> With three tymes as much of the fiery Dragon
> Mixing altogether, then hast thou well done.

mother and sisters, the thirteenth-century rhetor John of Garland discovers a schema for faculty psychology and Narcissus and Echo become formulaic sites representing reflection and reverberation.[9] I want to emphasize twelfth- and thirteenth-century science because it offers an established and even institutional approach to physical allegory. Too often we approach alchemy as a Renaissance occult science rather than seeing it as part of an established tradition of natural inquiry. The propensity to think in units so evident in alchemical allegory appears throughout the Renaissance in the use of adages, maxims, and aphorisms. Bacon's *Wisdom of the Ancients* is a well-known example:

> The truth is that in some of these fables, as well in the very frame and texture of the story as in the propriety of the names by which the persons that figure in it are distinguished, I find a conformity and connection with the thing signified, as close and so evident, that one cannot help believing such a signification to have been designed and mediated from the first, and purposely shadowed out (p. 404).

As Bacon demonstrated in the aphoristic style of the *Magna Instauratio*, such forms could easily be rearranged to promote new knowledge (Stephens, esp. pp. 98—136). The personal collection and arrangement of such narratives must not be overlooked. In contrast to the traditionally sponsored compendia, individual collections could take on special force. The Renaissance commonplace book represents a site for personal discovery and potential institutional subversion. Whether in moral philosophy or physics, the maxim or adage has the capacity to challenge or reassess a known body of knowledge.[10] In the sixteenth century, short narrative forms appear as rhetorical weapons against scholasticism and for opening scientific inquiry.

Early efforts to quantify alchemy deserve more attention. The best-known example is the *Monas Hieroglyphica* in which John Dee divides the alchemical process into 24 quasi-Euclidean theorems. Here fabulous narratives become transformed into chains of quasi-mathematical statements that follow Dee's earlier work with Euclid and with astronomy. While mythological material remains, its formulaic structure begins to be replaced by a geometrical taxonomy: "Theorem I. The first and most simple manifestation and representation of things, non-existent as well as latent in the folds of Nature, happened by means of straight line and circle."[11] In contrast to the elaborate and dense documents in Ashmole's *Theatrum Chemicum Britannicum* (1652) or Jennis' *Musaeum*

II

While historians of science often celebrate the appearance of atomistic theories and the notational systems that provide groundwork for modern chemistry, we should not ignore the efforts made to bring greater stability to alchemical narratives in the Renaissance. The oft-encountered claim that a certain text will truly convey the means of transmutation is evidence that practitioners desired stability. The need to repeat such declarations — to begin over and over again — marks, as Kuhn has suggested, the prescientific quality of alchemy (pp. 15—16). But it also tells us how conscious alchemists were about the absence of an experimental program. Evidence of such awareness abounds. The historical criticism or warnings against misleading alchemical books incorporated into many Renaissance works also marks an effort to find stability. Thomas Norton complains of those who make their books "full derke,/In Poyses, Parables, and in Metaphors alsoe,/which to Shollers causeth peine and woe" (Ashmole, p. 8). The author of *Bloomefield's Blossoms* portrays lost alchemists as "disguised Philosophers leane,/With Porpheries, and Morters ready to grinde and stampe,/Their heads shaking, their hands full of Crampe" (Ashmole, p. 308). Edward Kelly's prose is more direct:

Many books have been written on the art of Alchemy, which, by the multiplicity of their allegories, riddles, and parables, bewilder and confound all earnest students; and the cause of this confusion is the vast number and variety of names, which all signify and do set forth one and the same thing. For this reason I have resolved in my own mind to loosen and untie all the difficult knots of the ancient Sages (p. 113).

The Pharmacopoeias or pharmaceutical handbooks that began to appear at the end of the sixteenth century offer another example.[8] Because of their effort to reduce complex narratives into emblematic configurations, iconographic manuals such as Michael Maier's *Atalanta Fugiens* (1618) may also be regarded as instruments of reform.

An important reform closely related to fabulous narrative appears in the expanded use of short narrative forms. Physical allegory had long been associated with the fables incorporated into alchemical narratives. In the Middle Ages fabulous narrative became a learned instrument of natural philosophy precisely because of its formulaic narrative structure. For example, in the story of Pentheus' dismemberment by his

be regarded as alchemical documents. In other words, at the same time that the formulation of a universal vocabulary brings about a crucial alteration in the practice of chemistry, the process continues to be narrativized.

The narrative response challenges us to think further. The formation of a new notational system does not mean that narratives are abruptly removed or simply relegated to an antiquarian heap of pseudo-science. The Webster-Ward debate illustrates the desire to formulate a more useful chemical alphabet, but does not shift discussion away from narrative. For Webster — who argues that reference must be sought in the cabala — the appeal of universal notation is inscribed in *prisca scientia* (Debus, 1970, p. 39f.). The elemental language stands not as a pragmatic code for approaching nature but rather as the very language of God. The debate is important because it marks a shift toward narrativizing the discussion of language itself. Rather than concentrating on the interpretive possibilities provoked by fabulous narrative, analysis becomes directed at the notational system itself. Here we must make a critical distinction as well. While the analytical interest in notational systems becomes distinguished from fable or plot on the level of the local narrative, it remains closely related to the metanarrative. As plot or fable become criticized, the criticism itself reveals its own emplotted nature. Galileo's pronouncement that the book of nature is written in mathematical language offers an illustration (pp. 237—8). At the same time that it offers an implicit criticism of natural philosophy grounded in ordinary language, it reasserts a logocentric assumption through the metaphoric appropriation of mathematics in the book of nature. While mathematics offers a new language, it is envisioned in the context of an old book. The metaphor not only reminds us of Galileo's accomplishments, but testifies that nature's language will now be accessible only to a specialized group of mathematicians. The elitism associated with metaphysical narratives now transfers itself to a non-linguistic notational system. Where the power of interpretation once rested in a linguistic metanarrative, it now begins to shift to the notational system itself. Nevertheless, ordinary language continues to supply the context for applying a notational language such as algebra and for thinking about the evolving chemical codes. At the same time that an efficient, atomistic system of notation evolves, ordinary language remains not only the vehicle for application and commentary but the very field in which chemistry becomes defined.

In my concluding comments I want to suggest that even though early

Hermeticum (1625), Dee's *Monas* offers a simplification. In his effort to quantify alchemy, Dee anticipates the subsequent quantification of an atomistic theory of elements. While Dee's text remains impressive from the vantage point of organization, it is utterly arcane. The only master narrative to which it responds is Dee's own mental narrative.

While Dee seeks to reduce multiple texts into a single geometric system, the most frequent response was to undertake a systematic study of multiple texts. The exchange tables or code books mentioned above comprise an early example. But by far the most ambitious efforts appear later in the seventeenth century. Here it is above all Newton's elaborate effort to synthesize an ocean of alchemical texts that demonstrates how alchemy had come to depend on a working assumption of intertextuality.[12] Just as euhemerism informs Newton's work on history and anagogical allegory provides the code for his reading of Daniel and Revelations, physical allegory controls his approach to alchemical manuscripts.[13] Newton never relies on a single text but uses individual narratives as correctives for each other. For Newton, alchemy begins with reading and synthesis. Newton's 'rereading' — like Milton's rereading of the Bible and the early church fathers — carries with it the possibility of an entirely new master code.

III

Today we acknowledge that the emergence of a new code or system of nomenclature follows a reevaluation of narrative. Letters rather than plot become the means for more exacting investigation. As Boyle has it: "that all elemented bodies be compounded of the same number of Elements, [as] for a language; that all its words should consist of the same number of Letters" (p. 346). But the impulse to narrative is not so easily removed. Debus has shown how they continue well into the eighteenth century in mainstream chemistry.[14] Jacob Tollius' *Fortuita*, published the same year as Newton's *Principia* (1687), argued that the true meaning of ancient fables was related to chemistry. Aware of Newton's own intense occupation with alchemical research, Tollius' book becomes not simply a curious irony of the new juxtaposed with the old, but a means of approaching a vocabulary thoroughly familiar to the English scientist. At the height of the Enlightenment, *Les Fables égyptiennes et grecques devoilée* (1758) — it went through four editions — by Abbé Antoine-Joseph Pernety argued that ancient myths should

tives remain present. Others appear through comparisons of science and religion or in the social metaphors used to describe the cooperation of chemists in their new science. In their reverence for their discipline, Lavoisier would have chemists demonstrate a respect analogous to a culture's regard for literature: "Love of literary propriety has ceded ... to the love for science ... we have sought to imbue ourselves completely with this spirit".[17] Anderson underscores the importance of this relationship by noticing that here Lavoisier looks upon the individual scientist as part of an enterprise larger than his own interests, an enterprise he must accept as an act of faith. Elsewhere the spiritual metaphors that signal the accompanying narratives become even more social. For Macquer the accompanying narratives become even more social. For Macquer the clarity of a chemist's language — in contrast to the antisocial obscurity of the alchemist — designated that he was a good citizen: "These true citizens [of] chemistry would be able to provide excellent remedies ...".[18] In Lavoisier the social metaphor used to describe the evolving discipline are closer to the "esprit de corps" of a unified body of citizens or even a military force. "Chemistry marches towards its goal and towards its perfection by dividing, subdividing, and re-subdividing yet again; and we cannot tell where such successes shall end".[19] In contrast to the alchemist's belief in individual perfection, accomplishment for the new chemist becomes less dedication to an isolated search than adherence to a community vision. Where alchemy's goal rested in the projected negotiation of a complex process, chemistry's object becomes analysis of complexity; the discipline becomes characterized not by composition but decomposition. While the object of the chemist's enterprise shifts direction in comparison with the alchemist, the metaphors used to imagine its purpose continue to describe a belief in perfectibility.

CONCLUSION

My comments have concerned the function of narrative forms in a limited number of alchemical discussions from the Renaissance and chemical treatises from the eighteenth century. Initially I argued that alchemical texts should be approached by attending both to a local narrative (concerned with a specific laboratory function) and to a more abstract narrative that I have referred to as the metanarrative. While the metanarrative permits general social assent, the component or local

chemistry challenges alchemy's use of fable on the level of local narrative, it inherits aspects of the alchemical metanarrative concerned with purification. In effect the alchemical object of a psycho-physical purity becomes directed away from physical reality and toward the purification of language. As we saw above, concern with the obscurity of alchemical language is hardly a discovery of the new chemistry. Until the seventeenth century, however, such concern remains on the level of interpretation rather than representation. With critics like Pierre Joseph Macquer, representation becomes identified as a program in itself. As Anderson's important discussion shows, Macquer's *Dictionnaire de chymie* (1766) amounts to a manifesto directed not toward the matter of chemistry but toward the posture of knowledge.[15] For Macquer, chemistry's object is the cure and prevention of the alchemical disease that plagues language: "The Panacea, although certainly the maddest of all the ideas that entered into the heads of the alchemists, was however that which led to the founding of rational chemistry, and to its being raised upon the ruins of alchemy."[16] In essence the chemist is a doctor who finds in language both disease and cure. While Macquer's preoccupation with purification challenges the obscurity of alchemy's fables, it also reasserts alchemy's own ideal. It does so, however, by aligning alchemy's metanarrative with linguistic purification. Although applied examples of Macquer's manifesto appear throughout the eighteenth century, I have room to include only several from a major work on chemistry and language from the end of that century.

In his *Method of Chemical Nomenclature* [*Méthode de nomenclature chimique*] (1787) Lavoisier writes: "It is time to rid chemistry of obstacles of every kind which retard its progress and to introduce in it a true spirit of analysis; we have proved sufficiently that this reform must be brought about by perfecting the language" (Crosland, p. 131). Here the narrative once used to direct inquiry shifts from allegory to the metaphoric treatment of nomenclature itself. In a sense a new 'fable' is created that involves the purity of language. Once applied to the purification of natural elements, the process now occurs within language itself. The alchemical romance now appears as an effort to purify, cleanse, purge a diseased language. Lavoisier's comments are hardly unique but fit within an ongoing diagnosis of nomenclature in the eighteenth century. What we should notice is that dissolution of alchemical narratives hardly leads to the disappearance of narrative but to its concentration in metaphor.

Metaphors of disease, as we have noticed, become one way narra-

texts? At a time that metaphor supplies a means of opening comparisons between literary and scientific texts, it is necessary to question whether metaphor supplies sufficient epistemological ground for extended work. While our cultural understanding of scientific texts will certainly benefit from exploring the status of metaphor in regard to narrative, we also need to consider the ways it may constrict as well as enhance our inquiry.

NOTES

[1] For a study of the development of the scientific report, including the *Philosophical Transactions of the Royal Society*, see Bazerman.
[2] The work of Debus constitutes a major project in the study and reassessment of alchemy. Comprised of detailed histories, specific studies, bibliographic guides, and editions, his work not only challenges scholars to engage alchemy — and to undertake the thorough study of the discourse of chemistry in modern periods — but to learn from the ways early periods in the history of science were displaced by positivistic historians of science. For a recent meditation on the historiography of the history of science see Debus (1984).
[3] Sir George Ripley, *The Compound of Alchymie* in Ashmole (pp. 107—93). For an overview of the individual steps within the alchemical process see Shumaker (esp. pp. 170—73).
[4] For a discussion of Michael Maier, *Atalanta fugiens* (1618) and Basilius Valentinus, *Douze Clefs de la Philosophie* (1624) see J. Van Lennep.
[5] For a review of metanarrative in a discussion of legitimizing myths and narrative archetypes see Frederic Jameson's introduction to Lyotard (pp. vii—xxi); see also Jameson (1981).
[6] "We are confronted, then, from the outset and in a systematic fashion, with a double narrative, with two types of episodes, of a distinct nature but referring to the same event and alternating regularly ... the interpretation is included within the texture of the narrative. One half of the text deals with adventures, the other with the text which describes them. Text and metatext are brought into continuity" (Todorov, p. 123).
[7] For a discussion of hermeneutic instability with regard to fable see Knoespel (1985).
[8] The major work on alchemical semiotics remains Crosland. For discussion of pharmacopoeias beginning with the *Pharmacopoeia Augustana* (ca. 1564) see. p. 94f.
[9] Ghisalberti (lines 171—76; 163—66). For a discussion of Garland's commentary in the context of thirteenth-century physics see Knoespel (1985, pp. 38—45).
[10] Slaughter provides an even broader discussion of the implementation of axiomatic forms in the seventeenth century.
[11] Josten (pp. 84—221; p. 155); for Dee's use of language, see Knoespel (1987).
[12] "It was often Newton's custom in his note-taking to enter references to works other than the one being noted, presumably when it seemed to him that the other works expressed similar alchemical ideas. As he read more and more widely, his alchemical manuscripts were frequently more and more copiously annotated" (Dobbs, p. 130). For a recent discussion of Newton's reading habits see Westfall.

narratives are problematic sites and sources of obscurity. In the second part of my comments I noticed that the obscurity of alchemical narratives provoked a continuous effort to enforce a kind of hermeneutic stability in alchemy and contributed to the development of the new chemistry. Finally, I suggested that the hermeneutic stability associated with the new chemistry comes not from the wholesale dissolution of the alchemical narratives. Stability actually emerges from the extension of the alchemical metanarrative concerning purification to chemistry. Such a metanarrative, however, appears not in the form of a plot or fabula but asserts itself condensed in the metaphors of purification that accompany early chemistry.

At the beginning of this paper I suggested that the characterization of the transition from alchemy to chemistry as a paradigm shift greatly simplifies the complex transition. At this point I would add that such a formulation not only simplifies but engenders a blindness that comes from expecting shifts rather than continuity. My intention in this paper has been to suggest that the complexity of this transition may be approached through ongoing study of the narrative forms that contribute to both disciplines. In conclusion, I want to emphasize how the study of narrative may contribute even more to issues that have only been touched upon in this paper. My closing observations may be put in the form of questions:

1. *What relation do literary genres have to scientific narratives?* In regard to alchemy it is necessary to explore even more closely how romance narratives — characterized by Todorov as including both adventure and description of meaning — may be applied to alchemy. The object of such investigation would be found not in the simple alignment of literary and scientific texts but in the further exploration of ways the mythos of romance supplies the dominant narrative form in science.

2. *What role do short narrative forms have in the practice of science?* While much attention has been given to the larger forms that characterize science, the shorter narrative forms that accompany science have been the subject of little research. The story or word problem which has been part of scientific pedagogy and practice in all periods deserves special attention. In contrast to the conceptual shifts that often preoccupy attention in the history or philosophy of science, the short form displays a remarkable continuity.

3. *What relationship is there to metaphor and narrative in scientific*

Jameson, F.: *The Political Unconscious*, Cornell Univ. Press, Ithaca, 1981.

Josten, C.: 'A Translation of John Dee's "Monas Hieroglyphica" (Antwerp, 1564) with an Introduction and Annotations', *Ambix* **12** (1964) pp. 84—221.

Kelly, E.: *The Theatre of Terrestrial Astronomy* in *The Alchemical Writing of Edward Kelly* (ed. by A. Waite), Vincent Stuart & John M. Watkins, London, 1970 rpt.

Knoespel, K.: 'Fable and the Epistemology of Expanding Narrative', *University of Hartford: Studies in Literature* **17**:2 (1985) pp. 28—42.

Knoespel, K.: 'The Narrative Matter of Mathematics: John Dee's Preface to the *Elements* of Euclid of Megara (1570)', *Philological Quarterly* **66.1** (1987) pp. 27—46.

Knoespel, K.: *Narcissus and the Invention of Personal History*, Garland, New York, 1985.

Kuhn, T.: *The Structure of Scientific Revolutions* 2nd edn., Univ. of Chicago Press, Chicago, 1970.

Lyotard, J.: *The Postmodern Condition: A Report on Knowledge*, Univ. of Minnesota Press, Minneapolis, 1984.

Royal Society of London: Philosophical Transactions, Johnson Reprint Corporation, New York, 1963.

Shumaker, W.: *The Occult Sciences in the Renaissance*, Univ. of California Press, Berkeley, 1972.

Slaughter, M.: *Universal Languages and Scientific Taxonomy in the Seventeenth Century*, Cambridge Univ. Press, Cambridge, 1982.

Stephens, J.: *Francis Bacon and the Style of Science* Univ. of Chicago Press, Chicago, 1975.

Todorov, T.: *The Poetics of Prose*, Cornell Univ. Press, Ithaca, 1977.

Van Lennep, J.: *Art & Alchimie: Étude de l'Iconographie Hermétique et de ses Influences*, Editions Meddens, Bruxelles, 1966.

Westfall, R.: 'Newton and Alchemy', in *Occult and Scientific Mentalities in the Renaissance* (ed. by B. Vickers), Cambridge Univ. Press, 1984, pp. 315—35.

Georgia Institute of Technology

[13] An excentric but perceptive account of Newton's allegorical and anagogical practice is found in Castillejo (1981).

[14] My reference to eighteenth-century alchemical literature is indebted to Debus (1986), a paper delivered at a symposium on "Representation and Value: Literature, Philosophy, and Science." The symposium was jointly sponsored by The Program in Literature and Science in the Department of English at Georgia Tech and the NEH.

[15] My remarks on language in eighteenth-century chemistry are indebted to Anderson's excellent book.

[16] "La Médicine Universelle, quoique la plus folle sans doute de toutes idées qui étoient entrées dans la tête des Alchymistes, fut cepend qui commenca à établir la Chymie raisonnable, & à l'élever sur les ruines de l'alchymie" (Anderson, p. 32).

[17] "L'amour de la propriété littéraire a cédé ... à l'amour de la science ... nous avons cherché à nous pénétrer tous du même esprit" (Anderson, p. 123).

[18] "Ces vrais citoyens |de| la Chymie pouvoient fournir d'excellens remèdes ..." (Anderson, p. 32).

[19] "La chimie marche donc vers son but et vers sa perfection en divisant subdivisant, et resubdivisant encore, et nous ignorons quel sera le terme de ses succès" (Anderson, p. 136).

REFERENCES

Anderson, W.: *Between the Library and Laboratory*, Johns Hopkins Press, Baltimore, 1984.

Ashmole, E.: *Theatrum Chemicum Britannicum* (1652), Johnson Reprint Corporation, New York, 1967.

Bacon, F.: *Wisdom of the Ancients* (selections), in *Selected Writings of Francis Bacon* (ed. by H. Dick), Modern Library, New York, 1955, pp. 403–20.

Battaglia, S.: *La Coscienza Letteraria del Medioevo*, Editore Liguori, n.p., 1965.

Bazerman, C.: *Shaping Written Knowledge: The Genre and Activity of the Experimental Report*, Univ. of Wisconsin Press, Madison, 1988.

Boyle, R.: *The Sceptical Chymist* (1661), Dawson, London, 1965.

Castillejo, D.: *The Expanding Force in Newton's Cosmos*, Ediciones de Arte y Bibliofilia, Madrid, 1981.

Crosland, M.: *Historical Studies in the Language of Chemistry*, Dover, New York, 1978.

Debus, A.: 'Myth, Allegory, and Scientific Truth: An Alchemical Tradition in the Period of the Scientific Revolution' (Paper delivered at Georgia Institute of Technology, February 21, 1986).

Debus, A.: *Science and Education in the Seventeenth Century: The Webster-Ward Debate*, Macdonald, London, 1970.

Debus, A.: *Science and History: A Chemist's Appraisal*, Univ. of Coimbra, Coimbra, 1984.

Dobbs, B.: *The Foundations of Newton's Alchemy*, Cambridge Univ. Press, Cambridge, 1975.

Galileo: *The Assayer*, in *The Discoveries and Opinions of Galileo* (ed. by Stillman Drake), Doubleday, New York, 1957, pp. 229–278.

Ghisalberti, F., ed.: *Integumenta Ovidii: Poemetto inedito del secolo XIII*, Giuseppe Principato, Milan, 1933.

As the practical results of science have been progressively freeing us from arbitrary dictates of nature, they have also transferred nature's responsibility for those dictates to the human realm: each step into freedom from our dependency on nature brings with it the requirement to take on the responsibility for a course of action previously determined by forces of nature and to have conscious human decisions replace the enigma as well as the authority of nature's decrees.

The natural sciences, so it would seem, bring us the gift of freedom, but this gift comes without instructions, and, so it would seem further, those instructions would have to be supplied from elsewhere. The general tendency would appear to be directed toward locating this 'elsewhere' chronologically in an idealized past when traditional values had not yet been challenged and when social institutions entrusted with their preservation were regarded as the unquestioned arbiters of human affairs. The very question the editors of *Time* chose for their caption establishes the premise that once upon a time there were instructions or standards of conduct which contemporary society, to its grave detriment, no longer accepts.

This retrospective stance is made drastically evident by increasingly adamant attempts aimed at reestablishing prescientific value systems in order to avoid the moral challenge with which the progress of science has confronted our society. The most explicit recent example of this kind has been the politically supported attempt to introduce 'creationism' legally as a viable alternative to the scientific investigation of human origins. Had this attempt been successful, religious dogma would have effectively gained a position to claim not only equal status with science but also final authority for a prescientific moral code in a scientific age. As can be surmized from the general tenor of the debate and from the Supreme Court's decision, science was not primarily at issue but rather the moral context in which it is practiced. That this moral context was to be strictly circumscribed by Christian Fundamentalist doctrine, a doctrine that leaves little room for ambiguities, was, and is, undoubtedly the feature most attractive to supporters of the 'creationist' movement and its sympathizers.

My purpose in raising the matter of 'creationism' is not to hold its theoretical bias up for scorn nor to polemicize against it; my point is, rather, that science cannot be, and never is conducted in a moral vacuum and that the moral questions scientific endeavor raises cannot be dealt with independently of that endeavor itself. As a social phenom-

GÉZA VON MOLNÁR

"WHAT EVER HAPPENED TO ETHICS?"

On May 25, 1987, the cover page of *Time* bore the title "What Ever Happened to Ethics". The topic is timely, indeed; a good portion of the issue was devoted to it in an effort to survey a pressing social problem that has in recent times moved ever more forcefully into the foreground of public consciousness. The lead article bewails the current state of affairs by stating that "large sections of the nation's ethical roofing have been sagging badly, from the White House to churches, schools, industries, medical centers, law firms and stock brokerages — pressing down on the institutions and enterprises that make up the body and blood of America" (p. 26).

To this list we may well add the various disciplines in the natural sciences where research has fostered rapid and fundamental advances in human knowledge that remain unmatched by valid ethical criteria concerning the use to which such knowledge may be put. The comfortable pretension that no moral valence attaches to scientific objectivity has become ever more difficult to maintain over the years, and the progressivist faith in the inherent goodness of the goods science bestows on humankind has lost much of its appeal. In other words, awareness among scientists has grown with increasing urgency that they must ask themselves what their activities mean in the social context within which they conduct them, whereas members of the general public have come to realize that technological progress is far from being an unmixed blessing, for which reason it must be guided by standards that transcend the singular aim of exploiting nature and, instead, reflect the interests of the human community at large.

This growing recognition that the impact of science extends into the realm of public moral concern has made itself felt within our society in various ways; to name just a few of the most obvious manifestations that have become established features of the political forum, there are anti-nuclear movements, consumer interest groups, individuals and organizations devoted to the cause of ecology, as well as a host of other publicly voiced concerns similarly inspired but not organized to an equal degree.

Frederick Amrine (ed.) Literature and Science as Modes of Expression, 113–127.
© 1989 *Kluwer Academic Publishers, all rights reserved.*

renowned as a mathematician and philosopher; Kant, whose precritical lectures and publications dealt with a variety of scientific topics; Lichtenberg, who may be cited under the rubric of physics as well as literature; Goethe, the patriarch of German letters who pursued ambitious scientific projects; and Schiller, who, having received his educational training as a physician, gained lasting fame as a dramatist and derived important theories of history and literature from Kantian philosophy. Novalis, the name most intimately associated with German Romanticism, also belongs on this list. He is of special interest not only because he was equally active as philosopher, poet, and scientist — he was a geologist by profession — but also, or even more so, because the interrelatedness of these areas of endeavor receives its most explicit theoretical foundation in his work. Its distinguishing feature (in contrast to, say, Schelling's 'philosophy of nature') is the author's expertise as a practicing scientist who is familiar with scientific methodology and the results of current research in his field. In other words, Novalis manages to accommodate 'hard' scientific data with the most advanced developments in continental philosophy, which Kant had given a new, revolutionary direction. More than that, he himself may well be counted among those who contribute to these developments with his theory of poetics, or better yet his theory of language arts. As I shall attempt to outline, the link between theoretical reason, which gives rise to the realm of science, and practical reason, which governs the realm of ethics, becomes firmly established with the introduction of this theory of language or rather of '*Poesie*', to use Novalis' terminology.

With his First Critique, Kant redefined the relationship between self and nature that had been accepted as the epistemological framework relative to which things were constituted as objects of knowledge. The new framework still retained the interpolar stress between subjective and objective moments; however, the latter maintained its active contributions to the epistemological process only as the referential ground of experience, whereas the former lost its passive status to such a degree that a mimetic theory of knowledge would henceforth no longer be defensible. According to Kant, the object of knowledge is formed by the knowing agent's imagination in accordance with the spatio-temporal forms of sense perception and the categorical forms of instrumental reason, which furnish the context of all possible experience.

The point of Kant's redefinition of the operation of theoretical reason is not, as is still widely assumed, to subjectivize knowledge by

enon, 'creationism' would serve to illustrate this point; however, as an instance of social pathology, it also serves to illustrate that the codification of human worth and conduct is a historical process whose challenges cannot be met by appealing to the authority of precodified systems from the past.

This is not to say that the past cannot be instructive; it most certainly is, or so I shall argue. However, instead of our seeking to reestablish nostalgically perceived traditions, it would seem more helpful if we turned to the past with a different intent. Mine is to bring the relationship of interdependence between science and ethics into clearer focus. In doing so, I hope to furnish more than an exercise in philosophical archeology since it is this relationship that would seem to hold the key not only to our present dilemma but also to our perception of literature in its proper functional context.

"What ever happened to ethics?" might well be asked if we glance at our universities and research institutions where the natural sciences are pursued in categorical isolation from the social sciences and the humanities; there is even talk of the 'hard' sciences as opposed to the 'soft' ones, and the humanities are basically regarded as unscientific, even though some of the practitioners in that field would wish to claim 'scientific' legitimacy. Ethics is a matter left to be dealt with by philosophy departments and related disciplines. Even the term 'science' has come to be identified almost exclusively with the natural sciences and, concomitantly, they are regarded as the methodical model for all cognitive endeavor. Ethics, or the literary arts, do not come under this model and would, therefore, seem to be entirely unrelated to the natural sciences. This was not always so; the chasm that seems to separate science from ethics developed in the course of the 19th century with the rise of positivism.[1] Evidently, the present 'moral crisis' referred to by *Time* would indicate that it needs to be bridged, or rather undone, which can best be accomplished by first gaining some insight into the fundamental interrelation between science, ethics, and the literary arts as it was conceived before the chasm arose.

In the tradition of the eighteenth century, philosophy, science, and the literary arts represented interrelated rather than categorically separate fields of endeavor. Some of the most notable names associated with the period are not representative for accomplishments limited to one field alone. This is especially apparent in the annals of German history where one need only think of Leibniz, who was equally

individual's existence, condition it, and decree its demise are still to be regarded as the proper objects of scientific endeavor; only, the knowledge gained in all such endeavor is not neutral, not knowledge for its own sake, but knowledge whose significance as truth is one that may be acted upon and thus always carries an ethical valence. Even though Kant lays bare the interconnectedness that prevails between the spheres of theoretical and practical reason, he still treats each separately in his First and Second *Critiques* respectively, and it is actually Fichte who, in his *Science of Knowledge* [*Wissenschaftslehre*], fully outlines the hermeneutic circle that contains both. As moral philosophers they succeeded in establishing ethics as the common ground with reference to which all possible human knowledge attains its intersubjective significance. The large field of communicative intersubjectivity opened up in this context remains largely unexplored, however, and it is just this field toward which Novalis directs his attention.

When Friedrich von Hardenberg chose to enter the public realm of European Letters under a name derived from his ancestral title, 'de Novali', it was undoubtedly an apt choice since he clearly offered his contributions to the *Athenäum* in the same pioneering spirit that occasioned the brothers Schlegel to launch this, their journal in 1798. He and his friends were, indeed, clearing new land; they were the *avant garde* devoted to the self-assigned task of staking out new territory for literary theory and practice. Ever since those first forays, the name 'Novalis' would, sooner or later, appear on the banners of subsequent advances into new territory, even if, at times, such advances were merely putative and actually turned out to be retreats. At any rate, Hardenberg's choice of penname proved more valid then he could have expected since it became synonymous with Early German Romanticism and has been invoked nearly two centuries for various intellectual causes that were, rightly or wrongly, thought to be in the vanguard.

The diversity of factions who claimed Novalis as their own is quite extraordinary. The right accords him an honored place from the post-Napoleonic era of restoration to the period of Nazi ideology in our days, and the left, too, after prolonged silence interspersed with active disavowal, cites his name with increasing frequency and approval.[3] Catholics and Protestants have vied to enlist him in their ranks, and there are those who have created their own religious cult in his name;[4] moreover, believers in the religion of art, be they of pre- or post-Nietzschean denominations, would seem to entertain particularly strik-

relegating it to a realm that is effectively cut off from an objective sphere of reality, which supposedly contains things as they are in themselves. Rather the opposite is true. If we were able to know things in themselves, the thing would be as it is known and the act of knowing would be self-sufficient. This is exactly the premise on which the entire Western tradition of metaphysics rests since it deals with objects that are affirmed to be real without any means of reference to experience. As the history of that tradition would lead one to suspect, and as Kant demonstrates by example of his antinomies, reason exhausts itself in an empty game of dialectics unless it entails such reference to experience.

In redefining the operation of theoretical reason, Kant removed the basis for its being employed in a purely speculative manner, which is the manner of traditional metaphysics. However, this manner of employment had long been considered the highest of human achievements and had been pursued with great alacrity. That this pursuit was in error does not detract from the impulse that sponsored it. It had been misdirected and, speaking from a post-Freudian vantage point, the institutionalization of metaphysics may be regarded as the pathological symptom Kant's analysis was designed to treat. In the course of his analysis, it is revealed that the belief in our ability to know 'things in themselves' is fueled by an interest that is essentially practical and not theoretical.[2] This effectively means that the ultimate aim pursued in all cognitive acts is defined by practical, that is to say moral or ethical interest.

My reading of Kant's *Critique* would suggest that we are dealing with a text that differs from previous approaches to the philosophy of knowledge because it furnishes more than an analysis of faculties and functions involved in the process of cognition. Instead of an epistemology, Kant furnishes, in effect, the rudiments for a psychology of cognition. Reduced to a general formula, the *Critique* essentially states that we want to know what the world is, not in order to know 'the world', which in itself would be something entirely foreign and therefore unknowable in any case, but rather in order to determine its meaning with respect to human conduct. The interpretive or hermeneutic horizon for the constitution of all knowledge is thus the community of humankind as the ultimate referential ground for human agency.

Viewed from the perspective Kant offers in his philosophy, the multiform aspects of the natural environment that give rise to every

or the other domain exclusively. He does not speak for the worker or the cognitive model of the natural sciences, nor does he speak for any deity or the autonomy of art, nor yet for the ideal of a socio-political system that would correspond to the ambitions of the politically committed. What Novalis does speak for is a revised approach toward the understanding and appreciation of all human endeavor, a revision for which Kant's *Critiques* had laid the foundation and Fichte's *Wissenschaftslehre* had formulated the premise.

Kant's philosophy, as mediated by Fichte, essentially points out to potential recipients like Novalis that objects assume intersubjective validity only insofar as they may be acted upon in that same sense. The gains derived from this new perspective are, indeed, as Schlegel has claimed, revolutionary. It is a perspective from which the authority of the given is categorically put into question and laid open to denial, be that the authority of the object that would restrict reason to an instrumental function exclusively, or the authority of aesthetic standards embodied by works of art elevated to the rank of timeless models that would, consequently, have to remain unsurpassed, or the authority of socio-political institutions vouchsafed by norms rooted in the past that would make tradition the ultimate court of appeal to legitimate societal structures of organization and interaction. This challenge to authority opened up avenues for potential change in the natural sciences, the arts, and the socio-political arena. For various historical reasons, some of the avenues were traveled more successfully than others, but the promise of change in all theses areas of human endeavor was certainly felt and proclaimed by the *avant garde* spirits of the time, most notably by Novalis whose active concerns as a geologist, poet, and aware citizen extended across the entire spectrum of this promise.

If Novalis' efforts are not to be rendered out of context, it must be borne in mind that, for him, the negative momentum of challenging authority can arise only with reference to the positive momentum of relocating authority in the legitimating function of practical reason. Accordingly, he regards the universe of nature as a given text and all cognitive endeavor directed at it as a hermeneutic enterprise that falls within the horizon of practical reason and gains its intersubjective significance as language, or rather, as he would put it, as *Poesie*.

Novalis' concept of the poetic links the instrumental or theoretically determined function of reason to its free practical agency in a dialectic interplay that constantly engages both the determining moments of the

ing theoretical affinities with the author of *Heinrich von Ofterdingen* whose pronouncements on language, the arts, rationality, and various other related topics appear, in many respects, to be prefigurations of views held much later, down to those held in our day by poststructuralists (Molnár, 1987, p. 194ff.). He even had his following among the working class of a society that was moving toward industrialization since his poems dedicated to the craft of mining became songs popular with its practitioners, which may still be true in some instances.

Obviously that many conflicting views, attitudes, and interests could not possibly appeal to the same author with equal authority, and it seems just as obvious that the persistence of such willful appropriation cannot be satisfactorily explained in a manner more relevant to this appropriation's diversity, which may, to a certain degree, be accounted for in terms of historical change. It also seems quite futile to speak of misappropriation since it would prove difficult to base any such judgment on a standard that is not itself open to the same doubts and the same charge. The historically conditioned reasons and motives for appropriation must, it would appear, be granted their own respective merits along with the qualification that they tend to be mutually exclusive. Rather than fall prey to the same predicament by playing yet another round on the relativistic scale of appropriation and expropriation, I should like to suggest that different groups were receptive to Novalis because he did, in fact, address in the three distinct spheres of politics, aesthetics, and labor the entire spectrum of human interests, if not the specific manner in which those interests were variedly pursued.

Novalis' intellectual concerns and areas of engagement were many, yet they may well be categorized along the same lines that characterize the spheres of interest on the part of his audience: there are the pragmatic concerns, like those of the miners and of his own profession, which are addressed by the natural sciences; there are the aesthetic dimensions of human concerns, which are addressed by the arts, a task once shared by the mythopoetic tradition of religion, a tradition of which Novalis was keenly aware; and there are the socio-political concerns that raise intricate problems relative to human conduct, which must be addressed by determining the legitimacy or illegitimacy for societal norms. I should further like to suggest that Novalis regards those spheres, which correspond to the domains of theoretical reason, reflexive judgment, and practical reason, as inextricably linked, which would make any claim on him fallacious that is made in the name of one

tivism, the stage had been set for the estrangement that has come to be known as the division of society into 'two cultures', which diagnosis of our intellectual situation, accurate as it may have been, must, by now, be recognized as a numerical understatement.

Against this background of rationalism increasingly perceived without any reference to the non-instrumental capacity of reason, the arts were raised to champion freedom from subservience to reason altogether and to claim autonomy. However, there is no autonomy, except one derived from the principle of intersubjectivity, which is the non-instrumental principle of practical reason, and any claim of autonomy without reference to this principle is ill-founded. It is the same claim of autonomy made with an equal lack of foundation whether its sphere is limited to the cliquish aestheticism that, in its post-Nietzschean phase, proved all too susceptible to fascist ideologies of supremacy or whether it is expanded, under renewed Nietzschean impetus, to an all-inclusive supremacy that affects a revolutionary thrust more in tune with ideals of the left, as is the case with poststructuralism. Nietzsche had sought to define or rather unmask reason as an instrument of power; however, Nietzsche, Derrida, and Foucault notwithstanding, the instrumental concept of reason, even as an instrument of power raised to a neometaphysical level, simply misses the mark. All pretensions to the contrary aside, reason as instrument can only be the instrument of unreason and can, in this capacity, only introduce the negative momentum of heterogeneity, of radical relativism, which undermines any claim to priority advanced in advocating this position.

Yet, it would appear that the summary decryal of reason and Enlightenment tradition in our century arises from the same need or impetus that characterized the late Enlightenment and Early Romanticism. Both parties insist on questioning the meaning of knowledge. What Transcendental philosophers battled by laying open the unexamined premise of pre-Kantian dogmatism, right as well as left Nietzscheans battle by unmasking the tradition of Enlightenment rationalism, or rather Modernism, as a neo-dogmatism whose meaning is power. The right reacts to this insight with attempts to re-root society on pre-Enlightenment ground in order to cancel the Enlightenment's divisive effects on tradition; the left takes up the battle in order to free itself from the oppressive sterility of ideologies, systems, hierarchies etc. which were erected in the name of reason but failed to deliver on its promises. However, whereas Nietzscheans of the right remain radically opposed to the power of enlightened reason and would do away with human intellect in the ongoing process of communication. As distinct from our habituated manner of communicating, the poetic employment of language tends to emphasize the moment of freedom inherent in the very nature of communicative action. It is this characteristic of freedom raised to full consciousness in Romantic and post-Romantic practice that has given rise to claims of autonomy for the sphere of art and of the language arts in particular. However, art, language, or literature may be considered autonomous only on the strength of an appeal to the domain of practical reason, the domain of a communal ethos that refers community to human communality. Novalis most certainly never viewed artistic or linguistic autonomy in any context other than this one, that is he never lost sight of the ethical dimension freely introduced by practical reason as the primary and ultimate context within which human intersubjectivity may arise. Without reference to this non-instrumental feature of human reason, the sphere of language and literature would lack any basis for the mutuality that could allow for discourse or dialogue, indeed for the very possibility of any and all communication, other than its reduction to an exercise in solipsism that would assume the form of multiple soliloquies.

It is rather easy to overlook the practical province of reason and regard rationality only as an instrumental function since it is the instrumental capacity of reason that must be employed in order to cope with the technical aspects of executing practical commands in the realm of experiential reality. Once practical and theoretical reason are conflated into a concept of rationality that is dominated by the latter and can, therefore, be said to fulfill only an instrumental function, the stage is set for the new kind of objectivism that has characterized our civilization since the last century (Habermas, 1978). This objectivism is an entirely appropriate mental stance for the pursuit of knowledge in the natural sciences since it corresponds to the interests that constitute the practical horizon for that pursuit. Consequently, the dramatic progress achieved in the natural sciences supplies the added emphasis of obvious success that led to the general acceptance of scientific methodology as the model of cognition capable of setting standards for all areas of human endeavor. Since this model derives from interests limited to the communal effort aimed at mastering the given environment on which our physical being depends, it could prove only partially adequate for the social sciences and even less satisfactory for the humanities. In other words, with the entrenchment of this new objec-

objective validity comprises not things in themselves constitutive of 'nature,' nor merely the knowing self conscious of itself as a potential agent but all selves equally conscious. As knowing subjects we presuppose this collective self-referentiality and act upon it continually insofar as we attach to everything known or knowable a communal valence made manifest in the form of language, the medium of human communication. In its usual employment, however, the self-reflexive nature of language does not become apparent because, like the cognitive states it conveys, it seems to refer to an extra-cognitive realm of things. If language is to convey not only cognitive content but also the self-reflexive nature of such context, it must be employed in a manner that makes its own self-reflexivity evident. It is precisely this function Novalis ascribes to the poetic use of language and in order to employ it consciously in this manner one must also be able to behold the objective continuum of nature, the world, the cosmos, poetically. The ability to do so is equivalent to relating to nature or the world from the perspective of freedom granted by practical reason as outlined in Kant's and Fichte's philosophies, and it is this added perspective Novalis' *"höhere Wissenschaftslehre"* opens up.

From the vantage point afforded by the *"höhere Wissenschaftslehre,"* Novalis is able to arrive at a theory of poetics that allows him to assign language arts their indispensable function within the total spectrum of human endeavor. That spectrum entails three major categories, which may be refined or amplified as needed, but could be maintained in their rudimentary form for contemporary purposes of philosophical analysis. The field of endeavor claimed by theoretical reason comprises the natural sciences and is fueled by an interest in having nature accommodate human want and intent; this humanly significant factor in the acquisition of knowledge inspires the field of communicative interaction, that is to say the hermeneutic or language arts, and both, the sciences as well as language, are grounded in the human potential for ethical conduct which refers to an inclusive sphere of intersubjectivity and guarantees intersubjective validity for knowledge and its communicability. Effectively, the interrelatedness of science, language, and human conduct arises for Novalis within a self-reflexive continuum where the sciences seem to refer to nature but actually point to the question of intersubjective valence, acts of communication seem to refer to concepts but actually refer them to the intersubjective context of human communality, communal organization seems to refer to

the gains Western society owed to it, those of the left stand firmly on the political ground the tradition of the Enlightenment has prepared.[5] The paradoxical phenomenon of poststructuralists claiming on political grounds what they decry philosophically, namely the priority of reason over power and hermetic pantextuality, seems to indicate that their philosophical premises stand in need of reexamination. It also seems to me that such a reexamination would ultimately have to focus anew on the relationship of communication and communicability to ethics and to the natural sciences. In this respect, I believe, Novalis can once again be called upon to join the ranks of the *avant garde*.

Before he became productive as an author, Novalis engaged in an intensive discourse with Fichte's philosophy. The general trend of this discourse, as I read it, tends toward the establishment of a higher form of the *Science of Knowledge*, a "*höhere Wissenschaftslehre*", as Novalis states in his fragments of 1797/98. With it he is able to reach a perspective from which he can determine the function of language as essential in mediating, realizing, and recognizing the inherent but hidden relationship between cognition and ethics that Kant had outlined and Fichte had more stringently confirmed. Novalis' "*höhere Wissenshaftslehre*" differs from Fichte's in that its adds an explicitly theoretical dimension to practical reason, which for Kant as well as for Fichte is exclusively directed toward a free enactment of the moral imperative.

Practical reason is free because the moral imperative it enacts is not conditioned by needs and desires arising for any given self determined, as it is, by its relative spatio-temporal environmental setting. As a determined being, each individual can only act on self-interest, whereas the moral imperative identifies the interest of all selves as one overriding self-interest and thus enables every individual to join the family of humankind. It is this, the inherently ethical nature of practical reason, that furnishes the basis for intersubjectivity; however, to recognize it as such is another matter because the individual in its necessarily determined state ascribes this basis to that which is outside the individual and thus not individual, namely to its determinants collectively referred to as nature. That point of view is perfectly sound and legitimate, unless its narrow scope is considered closed, final, and absolute, which would prevent us from recognizing that the legitimacy of this point of view is not inherent in it but must be bestowed on it as collectively valid by the same mind that accepts it as such.

The referential context for any and all knowledge with claim to knowledge in the sciences, briefly but clearly recognized, was forgotten since progress in our knowledge of nature does not require philosophical reflection that links the very act of acquiring knowledge to an underlying common human interest and its ethical implications. Or does it? Has our neglect come to haunt us, not only with respect to the sciences but in all other areas of human endeavor, which a thinker like Novalis insisted must all be 'poeticized' — that is to say, their ethical grounding must be made evident? Is it possible our present dilemma proves his point? These are some questions, it seems to me, that may legitimately be raised, not in order to seek refuge in the past, which would be tantamount to advising the silliest form of escapism, but rather in order to retrace our steps, see where we left a potentially promising path for a less promising one, and gain a source of reorientation that may help us to deal with our present sense of loss and disorientation.[9]

NOTES

[1] For a detailed history of the rise and effect of positivism see Habermas, 1978, particularly p. 65 ff.

[2] This becomes explicitly evident in the "Methodological Part" ["Methodenlehre"] of the *Critique of Pure Reason*.

[3] Lukács has probably been most effective in tracing the intimate ideological links that lead from the Romantics to fascism; his views certainly appear to have been underwritten by the reception Nazis and their sympathizers accorded the Romantic School, which they also referred to as the "German Movement". For this and later variants in the history of reception, consult Peter.

[4] In his *Romantische Schule*, Heinrich Heine depicts one such follower, a young woman, with touching irony. For this as well as the other aspects of 'Novalismus', consult Albertsen and much of the secondary literature on Novalis that has been influenced by Rudolf Steiner and his school.

[5] In *Der philosophische Diskurs der Moderne*, Habermas argues along similar lines.

[6] C. S. Peirce may be cited as the first to recognize the communicative and communal foundation for the validation of scientific data, whereas Apel takes Peirce's position beyond the limits of scientific endeavor by defining the inclusive idea of intra-human 'consensus' as the necessary operative horizon for the possibility of any and all communication along with its attendant cognitive and practical dimensions.

[7] For substantiation of this claim, see my articles in *Goethe Jahrbuch* **96** (1979) pp. 270—279; *PEGS* **51** (1982) pp. 48—80; *Lessing Yearbook*, **14** (1982) pp. 23—42; *Goethe Proceedings at Davis*, Camden House, Columbia, S.C., 1984, pp. 77—91; *Goethe Yearbook* **2** (1984) pp. 137—222.

[8] For a detailed discussion of Novalis' poetic theory and practice, see my book *Romantic Vision, Ethical Context*.

political expedience but actually requires an ethos that regulates human interaction according to the potential of intersubjective equality, and this potential, in turn, furnishes the validating ground not only for any claim knowledge may have on intersubjective significance but also for the mediating power of language through which intersubjective significance may be established as such.[6]

This schema of interrelation lends a unique quality to Novalis' poetic work, which is particularly evident in his major prose fragments, his *Disciples at Sais* [*Die Lehrlinge zu Sais*] and his *Henry of Ofterdingen* [*Heinrich von Ofterdingen*]. Both are essentially conscious demonstrations of the self-reflexivity of language in that they have this very function of language, its essential poeticity, as their subject. However, unlike other authors on the evolving literary scene that spans the interim between his and our times, he never isolates language, or rather the poetic, from the larger context of the self-reflexive movement that encompasses the sciences and requires the dimension of ethics as its indispensable grounding. The interrelatedness of ethics and poetics from a Kantian perspective had already been proclaimed in theory and practice by Schiller, and other recognized authors of that time, notably Goethe,[7] may also be cited in this respect. Novalis' inclusion of science as essential rather than ancillary to his literary productivity is exceptional, however, and must be regarded as its most original feature that stands unmatched by his contemporaries. The singularity of this accomplishment is not to be thought of as an aberration or eccentricity but rather as the full envisionment of the entire scope open to modern poetics from the novel perspective granted by the level of consciousness Kantian philosophy had made possible.[8]

The synthesis of science, poetics, and ethics Novalis professed was to remain unique. For the most part, those engaged in the fields of literature and philosophy either became ever more fascinated with the self-reflexivity of language to the exclusion of its ethical moorings or they tended to socialize theories of poetics and philosophy in an effort to gain overt ethical relevance for both. The natural sciences, on the other hand, went their own way and those who pursued it were made ever more confident by their gains, so that their method became paradigmatic for all other endeavor, be that in the hermeneutic or social sciences, where positivism, until recently, came to be dominant.

What ever happened to ethics? It was lost on the way from the eighteenth century to ours. Its central function for the acquisition of

⁹ That this is possible is not that far-fetched if one considers the benefits philosophers of communicative and social theory, such as Karl-Otto Apel and Jürgen Habermas, have gained from precisely such a backward glance to the age of Kant.

REFERENCES

Albertsen, L.: 'Novalismus', *Germanisch-Romanische Monatsschrift* **48** (1967) pp. 272–85.
Apel, K.-O.: *Transformation der Philosophie*, Suhrkamp, Frankfurt, 1973.
Fichte, J.: *Science of Knowledge: with the First and Second Introductions*, Cambridge Univ. Press, Cambridge, 1982.
Habermas, J.: *Knowledge and Human Interests*, Heinemann, London, 1972 (rpt. 1978).
Habermas, J.: *Der philosophische Diskurs der Moderne*, Suhrkamp, Frankfurt, 1985.
Molnár, G.: *Romantic Vision, Ethical Context: Novalis and Artistic Autonomy*, Univ. of Minnesota Press, Minneapolis, 1987.
Peter, K.: *Romantikforschung seit 1945*, Hain, Königstein, 1980.

Northwestern University, Evanston

Having said that much about the unifying elements in romantic thought, we have to tend to the differences, which manifest themselves already at the level of the root metaphors. Organicism, as conceived by Coleridge, Wordsworth, Goethe, or Shelley, is, of course, a synthesizing metaphor: Coleridge speaks of a "balance or reconciliation" of opposites, while Wordsworth considers in the preface to the *Lyrical Ballads* "man and nature as essentially adapted to each other, and the mind of man as naturally the mirror of the fairest and most interesting properties of nature" (p. 738). A similar urge towards fusion is apparent in many of Novalis' chemical metaphors, which speak of mixing, amalgamation, and dissolution. Yet, as "dissolution" already indicates — and the older German term for chemistry, *Scheidekunst* (the "art of separating") loudly proclaims — the chemical metaphor is ambiguous. This is a complex issue that ought to be explored more; I shall use it now merely as a marker to raise the more general problem of unity, to which Géza von Molnár has devoted his latest, challenging study of Novalis. The traditional notion that Romanticism strove for unity, that it reached for a "reconciliation of opposite or discordant qualities," has been put into question by recent postmodernist criticism, a current that owes much of its own critical arsenal to its rereading of romantic texts. For the Anglo-American context, we may date the beginnings of this reinterpretation with the appearance of Paul de Man's now 'classic' article on "The Rhetoric of Temporality", which has become so familiar that I need not recapitulate its argument. Suffice it to note that although the second part of de Man's article contains a longer section on Friedrich Schlegel's concept of irony, postmodernist criticism has for a long time bypassed German Romanticism and Novalis in particular. Alice Kuzniar's recent book *Delayed Endings* on Hölderlin and Novalis is itself a delayed response to the new interpretive perspectives opened by postmodernism.

I see a number of problems in postmodernist reinterpretations of Novalis and I do not intend to apply it now to Novalis' conception of science. Nevertheless, I believe that postmodernist approaches, which usually pay little or no attention to romantic science, may occasion a reconsideration of older issues, and I should want to rely on them in developing my own view of Novalis' science, to which I shall now turn.

II

A consideration of 'nature as construct' may start with a note by

JOHN NEUBAUER

NATURE AS CONSTRUCT

I

A general consideration of Romanticism and science may start from the premise that all romantic poetics, however they may differ from each other, are metaphors taken from science or nature. The most influential among them describes both the working of the imagination and the structure of poems with metaphors taken from organic life: the imagination, as Coleridge writes in his *Biographia Literaria*, "reveals itself in the balance or reconciliation of opposite or discordant qualities" (ch. 14: 2, p. 16), and "the rules of the imagination are themselves the very powers of growth and production" (ch. 18: 2, p. 84). Goethe's similar notion of organicism is characterized by the concepts of type, metamorphosis, polarity, and enhancement [*Steigerung*]. Novalis deviates from this practice by choosing his metaphors not from organic life but, as Kapitza and Mahoney have shown, from chemistry, and as I have indicated in my book *Symbolismus und symbolische Logik*, from combinatorial mathematics.

Before we turn to these metaphoric organizations we should briefly consider what the implications of this consistent metaphoric 'scientification' (if you will excuse the term) of poetics are. It indicates that Romanticism, more than any other modern movement or epoch, defined its own poetic activity in relation to science and scientific perceptions of nature. It does not mean, of course, that the romantics valorized nature and science over poetry and poetic productivity. The reverse side of this metaphoric "scientification" was to conceive the practice of science as a fundamentally symbol- and metaphor-constructing activity: if poetics and poetry may be seen in terms of metaphors borrowed from science, science itself is a general 'poetic' activity because it perceives and organizes the sense-data by means of metaphors borrowed from elsewhere. Though these metaphors need not be borrowed from poetry, their constitutive role in knowledge makes science an imaginative enterprise. This, I believe, is the common romantic platform and the premise for their attempts to synthesize literature and science.

passage on Brown's medical theory. We may note, however, that such a control of nature is not without its problems, for it conjures up those dangers of an instrumental, exploitative, and tyrannical reason that Adorno and Horkheimer have studied in their *Dialectic of the Enlightenment.*

The quoted passage raises further questions, because Novalis, unlike Fichte, speaks of John Brown rather than of a transcendental consciousness and one may well ask whether Brown's theory has any intersubjective validity. Indeed, Novalis says that Brown's bold, imaginative, and constructive thought makes for a "true system" *only* for his followers. One must conclude that Brown's 'will to power' will inevitably clash not only with nature but also with the imperial drive and rhetorical power of other system builders. Such a Nietzschean-sociological approach to science is indeed adopted today by many philosophers of science (foremost among them Paul Feyerabend, partly even Thomas Kuhn), who see scientific theories as products of clashes between strong individuals, institutions, and institutional communities. Novalis' passage on Brown and his followers points toward such a sociological approach to science, but stops short of assigning a formative role to social forces in the formation of theory. We must therefore ask how, according to Novalis' modified Fichtean conception, agreements on scientific perceptions of nature may arise.

I suggest that we find not one but four, rather different answers to this question in his writings. Next to the sociological model, which is incommensurate with the tenor of Novalis' writings, we may identify three additional models in Novalis' philosophy of science. I shall label them (1) metaphysical, (2) transcendental, and (3) deconstructivist respectively. We may begin discussing them by turning to another *locus classicus* in his epistemology, the *Monolog.*

Speaking and writing are actually odd matters, true talk is mere word play. One cannot but be astonished by the ridiculous, mistaken belief that we are talking for the sake of objects. Nobody seems to understand that language is unique because it cares only about itself. This is why she is such a wonderful and fertile secret. If one speaks just for the sake of speaking one utters the most magnificent and original truths. But if we wish to speak about something definite, moody language forces us to say the most ridiculous and twisted things. That too is the source of the hatred that so many serious people feel against language. They notice her playfulness, do not notice however that her scorned chatter is the infinitely serious side of language. If only one could explain to people that language behaves like mathematical formulas — they form a world of their own, they

Novalis on the medical theory of a certain Scottish physician called John Brown:

> The best part of Brown's system is the astonishing confidence with which Brown presents his system as universally valid. It should be and must be so, experience and nature may say whatever they want. Therein lies the essence of every system, its truly effective force. Brown's system will thereby become the true system for his followers. Nothing can be said against this basically. The greater the magician, the more arbitrary his method, his utterances, his means. Everybody performs miracles *in his own way*.[1]

In earlier studies I have explored the historical connections between Brown and German Romanticism. But we need not know Brown's medical theory in order to recognize that Novalis attributes to it in this passage something that cannot be accomodated within the familiar concept of reconciliation with nature. The mind is conceived here as an imperial power that imposes its structure upon nature, or, better said, it constructs a nature where nature in a strict sense did not as yet exist.

Novalis' 'constructivist' approach to nature drew its foundation from Kant and Fichte. From Kant's "Copernican Revolution" he learned that the mind had an active, constructive role in ordering the world. Kant found mathematical constructions exemplary for this kind of structuring and Novalis' notes on Friedrich Murhard's *System der allgemeinen Grössenlehre* (1798) indicate that he understood this Kantian notion of construction and wanted to universalize it. When Murhard, in direct reference to Kant's *Critique of Pure Reason*, stated that the mathematician constructs, i.e. sensuously represents his concepts, Novalis commented: "Here too, it seems to me, the method of the mathematician is not individual. He plasticizes the concepts in order to fix them .... Why should the philosopher or, generally, every individual scientific master not do similarly? — One should plasticize spontaneously in all sciences".[2]

Fichte immensely broadened this constructivist freedom of consciousness, by interpreting both the self and the world as essentially constructions arising out of an originary act [*Tathandlung*] of positing. The formative role of this Fichtean idealism in Novalis' thought was traditionally, though often critically, acknowledged, and has been reaffirmed in recent studies by Hannelore Link, Richard Hannah, and Géza von Molnár. The latter, who devoted an earlier book already to Novalis' Fichte studies, eloquently and convincingly argues that there is a crucial ethical strain in Novalis' confrontation with the brute forces of nature — a confrontation for which we find an example in the quoted

skepticism concerns the possibility of ever achieving intersubjective agreement on the basis of moral imperatives.

Let me take a slight detour here. One may categorize Novalis interpretations by the way they interpret the variety in his thought, by their tendency to see either complementarities or incompatible alternatives in Novalis' different conceptions of nature. In an earlier article, which dealt with his poetics rather than his science, I characterized the anti-Fichtean alternative in Novalis' thought with Spinoza and pantheism. Crudely speaking: whereas Fichte advocated an overcoming of nature, Spinoza's pantheism meant — not only for Novalis but for all German romantics — a submission, a selfless merging and ultimate dissolution in the cosmos. The tendency towards a Spinozistic self-submission is manifest in every aspect of Novalis' life and work, for instance in his notion of love, his unique religious mysticism, his concept of disease, or his critical answer to the Fichtean concept of nature, succinctly summarized by the remark "statt Nicht Ich — Du" ["instead of not-I Thou"] (3, p. 430). I agree with Géza von Molnár that we ought to perceive this as a fundamental dimension of Novalis' thought, but we differ on its relation to the Fichtean component. Von Molnár interprets it as a "completion" of what he calls the Fichtean "basic schema" (p. 76 ff.), and he accomodates the self-assertive and self-submissive dimensions within some kind of complementary, reciprocating, and dialectical unity, which he sees as Novalis' ever-present goal, if not his achieved point of rest.

This is where we disagree and where postmodernism becomes relevant. While Novalis may have used the notion of an absolute, an overarching unity, as a kind of 'regulative principle', his life and thought are saturated with the fragmentary: it is manifest in his manner of writing, which yielded tentative, questioning notes, fragments (which he developed into a literary genre), unfinished novels, dialogues, and many other open forms. Von Molnár acknowledges these, but only as products created while striving for the absolute along a *via negativa*, while Alice Kuzniar reads them in her recent book as signs that Novalis realized and accepted radical temporality in de Man's sense. Neither von Molnár nor Kuzniar quote the first fragment from *Blüthenstaub* which says: "Wir suchen überall das Unbedingte, und finden immer nur Dinge" (2, p. 413). The sentence contains an untranslatable pun between *Dinge* [things] and *das Unbedingte* [the unconditional] and may therefore be translated only approximately as: "We search everywhere

play only with themselves, express nothing but their own wonderful nature, and for this very reason they are expressive, just because of this they mirror the strange interplay of objects.[3]

For our purposes it may suffice to single out the negation of the referentiality of verbal signs in this passage. Their release from a referential function endows them with "playfulness", and this may suggest that Novalis thinks of verbal signs like Jacques Derrida. Notice, however, that in our passage referentiality is reimported, so to speak, when the speaker of this monologue states that language becomes referential just when it is free. While individual words may not refer to anything in nature, combinations of words, the structure of language, "mirrors the strange interplay of objects". In contrast to the deconstructionists, who strip language of its metaphysical foundations, Novalis assumes here the existence of a metaphysical anchor.[4] It is this dimension of his epistemology that I call metaphysical, for it involves a model in which the correspondence between mind and world is guaranteed by something that is beyond both of them. That something is not so much a Cartesian God, who is a guarantor against epistemological skepsis, as a Leibnizian God, the origin of a preestablished harmony. On the basis of a preestablished harmony Novalis could at times even revert to the notion of natural signs: "The so-called arbitrary signifiers may in the end be less arbitrary than they seem — yet have a certain real relation [*Realnexus*] with the signified".[5]

Earlier critics have paid ample attention to this metaphysical dimension of Novalis' thought, which is deeply embedded in his theology and religion. More recent critics, myself included, would prefer an epistemology that is less dependent on such a metaphysical premise, but one must guard against reading one's wishes into Novalis' texts. Are there traces of a less metaphysical science in his writings? The major alternative that offers itself is transcendental philosophy, especially the Fichtean kind, where the problematic Kantian relationship between perception and nature is resolved by referring everything to processes in the mind. This is, as I understand it, von Molnár's line of thought, which, for all its attention to Novalis' concern with the absolute and his position in the history of mystical thinking, aims at an intersubjectivity that is founded on the subject and the community of subjects, and not on some metaphysical entity. I find von Molnár's attempt more appropriate than most recent interpretations, though I remain skeptical. My

ence at some other point. Within this "logical physics", experimentation must also become something different from laboratory work. What Novalis called experiment did not yet exist in his time but became a major form of scientific work in modern physics, namely 'thought experiment':

> Experimentation with images and concepts in the imagination quite in analogy to physical experimentation.[8] We shall become physicists only if we make *imaginative* materials and forces into a regulative standard of natural materials and forces.[9]

The key term in all these remarks is "experiment", or, better, mental experimentation. Consensus is nowhere postulated. While it is true that Novalis' manifest theory points towards a consensus, that his thought has a centripetal urge towards the center, I believe that there is an equally powerful centrifugal force operating in it, which consistently balances and often overweighs its counterpart. The monistic thrust is undermined by Novalis' pluralistic practice.

Von Molnár identifies Novalis' postulated consensus with *Poesie*: for Novalis, he writes, "theoretical reason, which brings world into consciousness, and practical reason, which guarantees intersubjective validation, are no longer linked by a circle of purely noumenal reality but by one immediately apparent to those who are aware of the 'mother tongue' the world speaks (Novalis, 1, p. 268). That awareness he calls *Poesie* and those who consciously employ language in this sense poets" (von Molnár, p. 133). In a more concrete and contemporary sense, von Molnár describes this 'mother tongue' as the language of "the ideal community of mankind", which acts as a normative regulative principle for all members:

> The unlimited community of communicants is the necessary ideal premise for every communicative act, and every such act is, at the same time, to be valued according to its relative contribution in the unbounded historical progress toward the realization of that community. . . . Since the normative regulative principle derives from the ideal community of mankind in which all relativism is suspended, it is fundamentally an ethical principle, so that a normative logic of science, which explains the world, presupposes not only a normative hermeneutic, which generates mutual understanding, but also a normative ethic" (p. 201).

Notice that in this explanation the intersubjective force of science comes not from some kind of agreement between signifier and signified (or systems thereof) but from an agreement between the communicants that is demanded of them by a "normative regulative principle". This

for the unconditional but find always only things". Von Molnár focuses on the first, Kuzniar on the second half of the remark, at the cost of doing justice to the other half.

III

How do these differences manifest themselves in the different interpretations of Novalis' concept of nature? None of the recent approaches address this question directly, although they answer it by implication. In my own view, which I had developed before the rise of postmodernism, we find a variety of approaches to nature in Novalis' writings and these do not cohere into an overall unity. This is perhaps most clear in his first fragmentary novel *Die Lehrlinge zu Sais* [*The Apprentices at Saïs*], where the second part contains a lengthy conversation between Fichtean, Schellingian, and other voices on nature, without coming to any kind of resolution (1, pp. 96—106).[6] This Bakhtinian 'dialogical' form is hardly accidental, and though one may of course argue that the differences would have been resolved in the sections that never materialized, one may also believe that incompletion was an intrinsic, if perhaps unintended, part of the design. Experimentation and exchange that appear here in the form of novelistic discourse were at the very heart of Novalis' concept of nature.

If nature is a construct for Novalis, this does not imply that it possesses a unique and binding configuration. Rather, it is a multiplicity of hypothetical alternatives existing side-by-side and changing in time. The construction of theory is no patient, systematic endeavor but rather an intuitive and ultimately artistic shaping of what is otherwise chaos:

The concepts of matter, phlogiston, oxygen, gas, force, etc. belong to a *logical physics* — that knows nothing of concrete materials — but reaches [*hineingreift*] idiosyncratically [*eigensinnig*] with bold hands into the world chaos — and creates an order of its own. Plotinus' physics.

Experimentation necessitates a genius for nature [Naturgenius], i.e. a wondrous sense to divine nature's sense — and to act in its spirit. The observer is an *artist* — he *guesses* what is *important* and is able to sense [herausfühlen] the important things among the strange mixture of phenomena that pass him by.[7]

Note the unusual meaning of "logical" in this passage. It does not refer to a causal enchainment of argumentation but rather to a mode of thinking that is purely mental, that does not take its point of departure from empirical observations — although it may call upon sense experi-

Hypothesizing is a risky game — in the end it becomes a passionate habit for untruth — and perhaps nothing has corrupted the best minds and science more than these bravadoes of the fanciful understanding. This scientific vice [*Unzucht*] dulls the sense for truth completely and weans away [*entwöhnt*] from strict observation, which, after all, is the only basis of all growth and discovery.
B: Hypotheses are nets, only those who cast will catch.[10]

Note, once more, Novalis' penchant for texts that speak with a plurality of voices: neither voice of the dialogue can unconditionally be attributed to Novalis, although one is, of course, inclined to see him as B rather than A. In any case, A is given a fair chance and his accusations are to be taken seriously.

If we take the plural voice of this dialogue seriously, we may perceive in it a recognition of, if not a plea for, methodological pluralism — a philosophical position that anticipates Paul Feyerabend's critique of methodological monism. If, however, we see A merely as a foil (something I am rather reluctant to do) we shall be closer to the methodological position of Karl Popper, who, in fact, has adopted B's reply as a motto for his seminal study *The Logic of Scientific Discovery*.

NOTES

[1] "Das Beste am Brownischen System ist die erstaunende Zuversicht, mit der Brown sein System, als allgemeingeltend, hinstellt — Es muß und soll so seyn — die Erfahrung und Natur mag sagen, was sie will. Darinn liegt denn doch das Wesentliche jedes Systems, seine wirklich geltende Kraft. Das Brownische System wird dadurch zum ächten System für die Brownianer. Dagegen läßt sich mit Grunde nichts mehr einwenden. Je größer der Magus, desto willkürlicher sein Verfahren, sein Spruch, sein Mittel. Jeder thut nach *seiner eigenen Art* Wunder" (2, pp. 545—6).

[2] "Auch hier ist das Verfahren d[es] Mathem[atikers] wie mich dünkt, nicht individuell. Er plastisirt die Begr[iffe] um sie zu fixiren und dadurch einen fest bezeichneten, sichren *Gang* und *Rückgang* nehmen zu können — Warum soll dies der Phil[osoph] nicht auch thun — oder überhaupt jeder einzelne wissenschaftliche Meister — In allen W[issenschaften] soll selbsthätig plastisirt werden" (3, p. 123).

[3] "Es ist eigentlich um das Sprechen und Schreiben eine närrische Sache; das rechte Gespräch ist ein bloßes Wortspiel. Der lächerliehe Irrthum ist nur zu bewundern, daß die Leute meinen — sie sprächen um der Dinge willen. Gerade das Eigenthümliche der Sprache, daß sie sich blos um sich selbst bekümmert, weiß keiner. Darum ist sie ein so wunderbares und fruchtbares Geheimniß, — daß wenn einer blos spricht, um zu sprechen, er gerade die herrlichsten, originellsten Wahrheiten ausspricht. Will er aber von etwas Bestimmten sprechen, so läßt ihn die launige Sprache das lächerlichste und verkehrteste Zeug sagen. Daraus entsteht auch der Haß, den so manche ernsthafte Leute gegen die Sprache haben. Sie merken ihren Muthwillen, merken aber nicht, daß das verächtliche Schwatzen die unendlich ernsthafte Seite der Sprache ist. Wenn man

leaves, as in Kant's ethics, a gaping hole between actual and ideal practice. One is inclined to object that science is no more a product of communicants striving for an ideal community than politics or business is the field where people are acting according to the Kantian categorical imperative. One may judge action by the standard of the categorical imperative but one cannot assume that it is *de facto* the standard that people use in deciding on action. In other words, it is a largely inoperative principle. To what extend Novalis was aware of this gap between what is and what ought to be, to what extent he thought it possible to proceed towards the consensus of dialogical clashes between irreconcilable opposites, is a matter on which there may be legitimate disagreements. But we must all admit that his notes, his fragments, his poetic works enact and juxtapose different voices or discourses that do not merge — no matter how much he wished consensus as an ideal. The rhetoric of this ideal should not overshadow the reality of his writerly practice.

IV

What is at stake here is not merely a matter that is internal to Novalis' writings and the kind of cohesion we attribute to them, but also matter that determines his historical place and his value for contemporary thought. For if the striving for unity and systematicity is, indeed, the central feature of Novalis' philosophy of science, then Novalis may well be grouped with Schelling and the *Naturphilosophen* whose intention was sometimes (though certainly not always) to *deduce* the laws of nature from unconditioned first principles. Hans Eichner's recent attack on romantic science does not do justice to Romanticism as a whole, but it rightly argues that this deductive *Naturphilosophie* runs counter to the main current of modern science.

My intent was to show that Novalis' notion of 'nature as construct' should not be confused with the construction of nature in *Naturphilosophie*, because it is much more experimental, hypothetical, and pluralistic. In all these respects Novalis' science is much closer to modern conceptions of science than to *Naturphilosophie*. We may consider in support of this conclusion a dialogue by him, which rehearses the arguments between an empiricist and a defender of hypotheses:

A: . . . A single faithfully observed fact is worth more than the most brilliant hypothesis.

Derrida, J.: 'La structure, le signe, et le jeu', in his *L'écriture et la différance*, Seuil, Paris, 1967, pp. 409—28.
Eichner, H.: 'The Rise of Modern Science and the Genesis of Romanticism', *PMLA* **97** (1982) pp. 8—30.
Feyerabend, P.: *Against Method*, Verso, London, 1975.
Gaier, U.: *Krumme Regel: Novalis' Konstruktionslehre des schaffenden Geistes und ihre Tradition*, Niemeyer, Tübingen, 1970.
Hannah, R.: *The Fichtean Dynamic of Novalis' Poetics*, Lang, Bern, 1981.
Kapitza, P.: *Die frühromantische Theorie der Mischung. Eine Untersuchung über den Zusammenhang von chemischer Wissenschaft und romantischer Philosophie und Dichtungstheorie*, Diss. Munich, 1968.
Kuhn, T.: *The Structure of Scientific Revolutions*, 2nd edn., Univ. of Chicago Press, Chicago, 1970.
Kuzniar, A.: *Delayed Endings. Nonclosure in Novalis and Hölderlin*, Univ. of Georgia Press, Athens, Ga., 1987.
Link, H.: *Abstraktion und Poesie im Werk des Novalis*, Kohlhammer, Stuttgart, 1971.
Mahoney, D.: *Die Poetisierung der Natur bei Novalis*, Bouvier, Bonn, 1980.
Molnár, G. von: *Novalis' Fichte Studies. The Foundations of his Aesthetics*, Mouton, The Hague, 1970.
Molnár, G. von: *Romantic Vision, Ethical Context. Novalis and Artistic Autonomy*, Univ. of Minnesota Press, Minneapolis, 1987.
Neubauer, J.: 'Dr. John Brown (1735—1788) and Early German Romanticism', *Journal of the History of Ideas* **28** (1967) pp. 367—82.
Neubauer, J.: *Bifocal Vision. Novalis' Philosophy of Nature and Disease*, Univ. of North Carolina Press, Chapel Hill, 1971.
Neubauer, J.: *Symbolismus und symbolische Logik. Die Idee der Ars Combinatoria in der Entwicklung der modernen Dichtung*, Fink, Munich, 1978.
Novalis, *Schriften* (ed. by P. Kluckhohn and R. Samuel), 4 vols., 2nd edn., Kohlhammer, Stuttgart, 1960—75.
Popper, K.: *The Logic of Scientific Discovery*, Basic Books, New York, 1959. 1st German edn. 1934.
Wordsworth, W.: *Poetical Works*, Oxfort Univ. Press, London, 1950.

Instituut voor algemene Literatuurwetenschap, Amsterdam

den Leuten nur begreiflich machen könnte, daß es mit der Sprache wie mit den mathematischen Formeln sei — Sie machen eine Welt für sich aus — Sie spielen nur mit sich selbst, drücken nichts als ihre wunderbare Natur aus, und eben darum sind sie so ausdrucksvoll — eben darum spiegelt sich in ihnen das seltsame Verhältnißspiel der Dinge" (2, p. 672).

[4] See in this respect Jacques Derrida's article "La structure, le signe, et le jeu", which distinguishes between a nostalgic play with signs of a structure that has lost its center and a Nietzschean free play of signifiers. Novalis' notion of play definitely belongs to the first category.

[5] "Die sog[enannten] willkührlichen Zeichen dürften am Ende nicht so will[kührlich] seyn, als sie scheinen — sondern dennoch in einem gewissen Realnexus mit dem Bezeichneten stehn" (3, p. 305).

[6] Gaier and Mahoney have shown that this dialogue is tightly organized, but they have not demonstrated that a consensus finally emerges.

[7] "Der Begriff Materie, Phlogiston, Oxigène, Gas, Kraft etc. gehören in eine *logische Physik* — die nichts von concreten Stoffen weis — sondern mit kühner Hand eigensinnig in das Weltchaos hineingreift — und eigne *Ordnungen* macht. *Plotins Physik*.

Zum Experimentiren gehört *Naturgenie*, d. ist, wunderartige Fähigkeit den Sinn der Natur zu treffen — und in ihrem Geiste zu handeln. Der ächte Beobachter ist *Künstler* — er *ahndet* das *Bedeutende* und weiß aus dem seltsamen, vorüberstreichenden Gemisch von Erscheinungen die Wichtigen herauszufühlen" (2, p. 179).

[8] "Experimentiren mit Bildern und Begriffen im Vorstell[ungs] V[ermögen] ganz auf eine dem phys[ikalischen] Experim[entiren] analoge Weise. Zus[ammen] Setzen. Entstehen lassen — etc" (3, p. 443).

[9] "Wir werden erst Physiker werden, wenn wir *imaginative* — Stoffe und Kr[äfte] zum regulat[iven] Maaßstab der Naturstoffe und Kr[äfte] machen" (3, p. 448)

[10] "— Eine einzige wahrhaft beobachtete Thatsache ist doch mehr werth, als die glänzendste Hypothese. Das Hypothesiren ist eine risquante Spielerey — Es wird am Ende Leidenschaftlicher Hang zur Unwahrheit — und vielleicht hat nichts den besten Köpfen und den Wissenschaften mehr geschadet, als diese Renommisterey des fantastischen Verstandes. Diese szientifische Unzucht stumpft den Sinn für Wahrheit gänzlich ab, und entwöhnt von strenger Beobachtung, welche doch allein die Basis aller Erweiterung und Entdeckung ist.

B. Hypothesen sind Netze, nur der wird fangen, der auswirft (2, p. 668).

REFERENCES

Adorno, T. and Horkheimer, M.: *Dialektik der Aufklärung*, S. Fischer, Frankfurt, 1971.
Bakhtin, M.: *The Dialogic Imagination*, Univ. of Texas Press, Austin, 1981.
Coleridge, S.: *Biographia Literaria* (ed. by J. Engell and W. J. Bate), Princeton Univ. Press, Princeton, 1983.
de Man, P.: 'The Rhetoric of Temporality', in his *Blindness and Insight: Essays in the Rhetoric of Contemporary Criticism*, 2nd rev. edn., Univ. of Minnesota Press, Minneapolis, 1983, pp. 187–228.

on associationism and faculty psychology. Following the Scottish school of mental philosophy, Rush posited nine basic capacities or "faculties" in the human mind, grouping these nine faculties into three categories: the "passions" included the passions *per se*, the will, and faith or "the believing faculty"; the "intellectual faculties" encompassed the reason or understanding, imagination, and memory; and the "moral faculties" included the moral faculty itself, conscience, and a sense of deity (Carlson, p. ix).[2] Insanity had long been recognized as a disease affecting what Rush called the intellectual faculties. Where Rush broke with traditional psychiatric theory was in declaring that insanity did not necessarily involve a disorder of the intellect, that the moral faculties alone were capable of succumbing to disease (Carlson, p. x). Like Philippe Pinel in France, he realized that a form of insanity might occur which perverted the sense of moral responsibility necessary to deter crime. Thus in a normal individual, an innate moral sense could stave off the passions while the intellect calmly concluded the proper conduct. But if this moral sense, this power to distinguish between good and evil, were momentarily suspended, the opportunity for calm inquiry would be denied, and the individual's will would become committed to a criminal act before his reason could repudiate it (Rush, 1972, p. 1). He would then become the victim of an "irresistible impulse" forced upon the will "through the instrumentality of the passions" (Rush, 1830, pp. 262; 355—57). In modern terminology, he would be emotionally disturbed.

Startling as it was, Rush's theory of "moral derangement" received little attention in America before the 1830s. Then, in 1835, James Cowles Prichard published his classic discussion of the problem in *Treatise on Insanity and Other Disorders Affecting the Mind*. This work popularized the study of what Prichard termed "moral insanity", making it, in the words of one historian, the "focus of psychological studies and polemical arguments until the end of the century" (Carlson, p. xi). Following the leads of Pinel and Rush, Prichard restated and developed the body of theory which would eventually lead to the classification of psychopathic personalities. He posited a disease in which

> the intellectual faculties appear to have sustained little or no injury, while the disorder is manifested principally or alone, in the state of the feelings, temper, or habits. In cases of this description the moral and active principles of the mind are strangely perverted and depraved; the power of self-government is lost or greatly impaired; and the individual is found to be incapable, not of talking or reasoning upon any subject proposed to him, for this he will often do with great shrewdness and volubility, but of

"OBSERVE HOW HEALTHILY — HOW CALMLY I CAN TELL YOU THE WHOLE STORY": MORAL INSANITY AND EDGAR ALLAN POE'S 'THE TELL-TALE HEART'

David R. Saliba has recently argued that Edgar Allan Poe's "structural omission of an objective viewpoint for the reader [in 'The Tell-Tale Heart'] forces the reader to experience the tale with no point of reference outside the framework of the story". "The reader", says Saliba, "is led through the story by the narrator with no sense of reality other than what the narrator has to say". This narrative technique forces the reader to identify with the narrator and to take the narrator's values as his own (pp. 142—43n). What Saliba fails to realize is that no one can read a text without an external sense of reality; all audiences bring to a work of literature some frame of reference that exists outside the text. And for Poe's audience in the 1840s, that frame of reference would have included a knowledge of a controversial new disease called 'moral insanity' and of the legal and philosophical dilemmas that surrounded its discovery. Poe's narrator in 'The Tell-Tale Heart' is a morally insane man, and Poe would have expected his readers to locate the symptoms of that condition in the language of his narration. Thus if we are to recover the meaning of the tale for Poe's audience, an audience that applauded 'The Tell-Tale Heart' at the same time that it shunned tales like 'Ligeia', 'William Wilson', and 'The Fall of the House of Usher' — indeed, if we are to assess the tale's significance for today's audience — we need to establish the medical history from which Poe drew.

We begin, then, with the 'father of American psychiatry', Benjamin Rush. In 1787, Rush was placed in charge of the insane at the Pennsylvania Hospital, and his work in this institution culminated in the first book on psychiatry by a native American, *Medical Inquiries and Observations upon the Diseases of the Mind* (1812). In his introduction to two of the essays Rush included in *Diseases of the Mind*,[1] E. T. Carlson explains how Rush developed a new theory of insanity based

Frederick Amrine (ed.) Literature and Science as Modes of Expression, 141—152.
© 1989 *Kluwer Academic Publishers, all rights reserved.*

But the concept of moral insanity changed all this, and the legal dilemma posed by this new definition of madness was obvious. If God had so constituted men that their passions or impulses were not always governable by an intact reason, how could society punish them for indulging in these passions? As pleas of moral insanity became increasingly common, this question stymied a criminal court system established as an instrument of retribution rather than as an agency for determining mental health.[4] A reaction was inevitable and almost immediate. Judges found themselves asserting that moral insanity was, in Baron Rolfe's words, "an extreme moral depravity not only perfectly consistent with legal responsibility, but such as legal responsibility is expressly invented to restrain" ('Baron Rolfe's Charge to the Jury', p. 214). Some of America's leading pre-Civil War psychiatrists — men like Isaac Ray, Samuel B. Woodward, and Amariah Brigham — wrote numerous treatises and periodical articles delineating the characteristics and supporting the pleas of criminals claiming moral insanity, but they faced serious opposition within their own profession almost from the outset, and by the late 1840s even some distinguished asylum superintendents began denying the existence of a 'moral' insanity.

The views of these skeptical physicians were generally more in keeping with public sentiment, and those medical men who supported an accused murderer's claim of insanity came under increasingly sharp attacks in the periodical press. The average man tended to suspect deception in defense pleas of insanity, and newspapers often fanned these feelings. Thus by the time Poe wrote 'The Tell-Tale Heart,' such trials were major events. When William Freeman was tried for the stabbing murders of the prominent Van Nest family in New York, the counsel included John Van Buren for the prosecution and ex-governor William H. Seward for the defense. Papers across the country kept track as seventy-two witnesses were called to testify as to his sanity, including a who's-who list of medical authorities, and Freeman himself, housed in a cage outside the courthouse, was the subject of "uncounted spectators" until he died of consumption in his cell almost eighteen months after his offense (*The Trial of William Freeman*, pp. 68—71, 79—80; Fosgate, pp. 409—14).[5]

Freeman and those like him were, to use the modern slang, "hot copy". The journals of the day devoted thousands of pages to analysis of them. Philosophical and literary societies debated the ethical and moral implications of decisions surrounding their cases. And writers

conducting himself with decency and propriety .... His wishes and inclinations, his attachments, his likings and dislikings have all undergone a morbid change, and this change appears to be the originating cause, or to lie at the foundations of any disturbance which the understanding itself may seem to have sustained, and even in some instances to form throughout the sole manifestation of the disease (pp. 4—5).

A disturbance of the emotions could be both the cause and the "sole manifestation" of mental illness. The morally insane man might be rational, might realize that those around him would condemn his behavior, but he himself would not.

In the decade following the appearance of Prichard's study, the concept of moral insanity became the topic of political, social, and theological debate both at home and abroad. As Rush himself foresaw, any new theories which emphasized the power of man's emotions to determine his actions occasioned intense hostility when they conflicted with other, presumably more agreeable, ideas about human nature. Such theories were opposed on the grounds that they degraded the quality of man's spiritual life, and for the more pragmatic reason that they reduced the incentives for good behavior. But nowhere were the new theories on moral insanity argued more strenuously that in the courts. Prior to the work of men like Rush and Prichard, if a person pleaded insanity in a court of law, he was presumed to be either an idiot or a raving maniac. A review of press releases concerning these trials, and of verbatim trial reports, shows that judges, counsel, witnesses, and observers tended to use three major criteria to establish insanity: the accused had to be unable to recognize right from wrong; he had to be illogical and virtually witless at all times; and he had to reveal a violent disposition before committing his offense ('Homicidal Insanity', p. 279; Wharton, I, 162—72). John Haslam's discussion of the jurisprudence of insanity in *Observations on Madness and Melancholy* (1810) reveals that madness was considered to be, in Haslam's words, as opposed to "reason and good sense as light is to darkness"; in order to exempt a man from criminal responsibility, the defense had to establish that he was "totally deprived of his understanding" and no more knew what he was doing "than an infant, than a brute, or a wild beast" (Haslam, 1975, p. 31; see also Coventry, p. 136). A man who, like Poe's narrator in 'The Black Cat', became unaccountably brutal, set fire to his home, and violently murdered his wife could not be judged insane if he appeared 'normal' to witnesses at the time of the trial.[3] And anyone who fled from the scene of a murder, or tried to hide the evidence, was legally sane because he was presumed to know right from wrong.

I had no desire. I think it was his eye! yes it was this! One of his eyes resembled that of a vulture — a pale blue eye with a film over it. Whenever it fell upon me, my blood ran cold; and so by degrees — very gradually — I made up my mind to take the life of the old man, and thus rid myself of the eye forever (1978, p. 792).

Poe skillfully refrains from divulging exactly what the narrator fears, and his readers have consistently picked up the gauntlet and put forth their own theories. Robert Shulman believes that the filmed-over eye suggests that the old man is cut off from "insight into the ideal and the beautiful" and that the narrator's fear thus represents man's "psychological dread that existence is meaningless", or more specifically, is a reflection of Poe's feelings toward the stepfather who "called into question the meaning of [his] life" (pp. 259—60). Arthur Robinson argues that the feared "Evil Eye" is actually the "Evil I", that the narrator "images himself as another and recoils from the vision" (pp. 101—2). And in his introduction to 'The Tell-Tale Heart', T. O. Mabbott concludes that the tale is founded on the "popular superstition" of the Evil Eye and points out that Poe may even be suggesting that it really *is* the old man's eye which drives the otherwise sane narrator mad (Poe, 1978, p. 789). However we feel about these interpretations, we should perhaps realize that much of Poe's audience, and certainly Poe himself, would have been familiar with Rush's theory (1830, p. 173) that the insane were "for the most part easily terrified, or composed, by the eye of a man who possesses his reason". They would have surmized that Poe's narrator is terrified by, in Rush's words, "the mild and steady eye" of a sane man.[7]

But it is not the eye alone which brings about the final decision to take the old man's life. Rather, it is a *peculiar sound*, and to understand the medical significance of this sound, we must go back to the beginning of the tale. The narrator opens his defense by declaring that although he is "very, very dreadfully nervous", he is not mad (1978, p. 792). Poe's readers probably would have recognized his nervousness as one of the common predisposing causes of moral insanity. Certainly most physicians writing at the time of Poe's tale would have agreed with Samuel B. Woodward (p. 288) that moral insanity, unlike mere depravity, was always preceded or accompanied by "some diseased function of the organs, more or less intimately connected with the nerves". Rush had maintained that "all those states of the body ... which are accompanied with preternatural irritability ... dispose to vice" (1972, p. 20). But even if the audience was uncertain about the

like Poe — who was himself a trial reporter in the 1843 murder-by-reason-of-moral-insanity trial of James Wood ('The Trial of James Wood', pp. 105—106)[6] — used them as models for some of their most disturbing creations.

One of these creations came to life in 'The Tell-Tale Heart'. Defendants in moral insanity trials were rarely allowed to speak in their own behalf, but Poe would let his character speak, and as he spoke, he would inadvertently let slip the very evidence which would establish him as morally insane.

The first thing we should notice about Poe's narrator is that his monologue is actually a long argument trying to establish not his innocence — he has already confessed to killing the old man — but rather his sanity. He builds this argument on the premise that insanity is irreconcilable with systematic action, and as evidence of his capacity for the latter, he explains how he has executed an atrocious crime with faultless precision. "This is the point", he tells us: "You fancy me mad. Madmen know nothing. But you should have seen *me*. You should have seen how wisely I proceeded — with what caution — with what foresight — with what dissimulation I went to work!" (Poe, 1978, p. 792). A madman, he implies, would be out of control, would be profoundly illogical and not even recognize the implications of his actions. His art in planning and coolness in executing his crime prove that he has the lucidity, control, and subtle reason which only a sane man could possess.

Poe's narrator is, of course, relying upon the old criteria used to establish insanity. But it would have been difficult for an audience reading his words in 1843 not to call to mind the medical publications and trial reports filling the popular press with a new theory of insanity. If they knew enough about this new theory, they might even have recognized Poe's narrator as a fair representation of Prichard's morally insane man. Like the patients in Prichard's study, he is capable of reasoning "with great shrewdness and volubility", but "his attachments . . . have undergone a morbid change" (Prichard, pp. 4—5).

This is not to say that Poe's narrator is always rational. He may be able to carry out his crime with a cool precision, but as he himself explains, his determination to murder his old friend stems from an irrational fear of his eye:

Object, there was none. I loved the old man. He had never given me insult. For his gold

readers, they contained new and fascinating information for Poe's. And this, of course, is the point. New medical theories were forcing upon Poe's audience questions of ethical moment and challenging their old ideas about the nature of man. It may even be that this audience, like most of the students I teach today, found the real terror in the story lay in identifying themselves not with the narrator, as Saliba suggests, but with the victim. It certainly would have been natural for Poe's 1843 readers to see themselves as the victims of the morally insane men discussed in the popular press, just as twentieth-century readers tend to associate themselves more with suffering families and felled presidents than with madmen who attack McDonald's and presidential assassins. In any case, Poe's narrator is maintaining a causal sequence — I can reason; therefore I am not insane — which Poe's audience had just discovered was false, so that it is not only the experiences the narrator reports that are unusual and problematic, but the report itself. "Observe how healthily — how calmly I can tell you the whole story", he begins (1978, p. 792). But "calmly" could no longer be equated with "healthily". The narrator's explanation fails to coincide with his audience's knowledge, and the implication is that Poe intends to display this disagreement in order that the audience might experience and evaluate it. Far from being trapped inside the story, the audience would stand outside the narrative and use its knowledge of the current medical controversy to replace the speaker's version of events with a better one, or even to question the moral implications of such an argument.

The narrator tells them that he has suffocated an old man because of his eye. But to make such an argument is finally to flaunt your lack of motive, and indeed he begins his explanation by admitting that "object there was none" (1978, p. 792). Those readers who insist upon positing an external motive on the narrator's part, or an unconscious motive on Poe's, deny the story some of its power. Like the murder of the Van Nest family, this murder is all the more terrifying because it is gratuitous. The narrator's obsessions have no logical object in the manifest text, and the tension produced by his explaining at length something for which there is no satisfactory explanation took Poe's story to the heart of the vexing question of moral responsibility as it dramatized the increasingly problematic nature of the human personality.

For Poe's 1843 audience, the new medical science had done more than just drag Diana from her car; it had questioned the integrity of even the 'rational' mind. But what about today's audience? Clearly the

significance of the narrator's dreadful nervousness, they certainly would not have been uncertain about the significance of his next statement. This nervousness, or "disease", had "sharpened [his] senses", he tells us, "not destroyed — not dulled them", and "above all was [his] sense of hearing acute" (1978, p. 792).

It would be difficult to think of a worse argument for sanity in 1843 than what Poe's narrator calls his "over acuteness of the senses " (1978, p. 795). Medical opinion at home and abroad had long held that "there is scarcely any symptom more frequently attendant upon maniacal ... disorders than a defect, excess, or some kind of derangement in the faculty of hearing" (Reid, p. 190), and that it is frequently "noises in the ear, such as sounds made during the night in the chimney", and in particular, "the noises of clocks and of bells" (Sigmond, p. 589; 'On Impulsive Insanity', p. 620) which haunt the minds of these men.[8] We should not be surprised then to learn that, as he stands over his intended victim, Poe's narrator hears "a low, dull, quick sound, such as a watch makes when enveloped in cotton" (1978, p. 795). His assertion that "he knew *that* sound well, too" reminds us that he has also been hearing another sound — that of the death-watches in the wall. "Night after night" he had listened to their ticking, telling himself that "it is nothing but the wind in the chimney", until the night when, "excited to uncontrollable terror" by the noise, he stalks his victim (1978, p. 796).

After the murder, the ticking sound returns, and the fear, outrage and paranoia it inspires increase until the seemingly rational murderer must confess his crime to the unsuspecting police. Even this confession would have been considered strong evidence of moral insanity. In a typical case from 1832, a man on trial for the murder of his son was found insane because he had "slaughtered his unoffending son to whom he should have been attached", and then confessed. One reporter explains:

The confession of the crime, I conceive, may be considered as an evidence of insanity of considerable weight. Not that every man who confesses a murder is to be considered insane, but, by this, taken along with other circumstances, as when the individual ... attempts to give reasons for the propriety of his conduct, we have a strong indication ... of the deranged condition of the intellect .... In short, it is so universal in such cases, that some very distinguished medical jurists consider this confession alone to be a significant test of insanity (Watson, 1832, p. 47).[9]

Observations such as this can be found throughout the trial reports of the 1830s and 40s, and while they may sound fairly obvious to today's

In the first chapter of *The Narrative of Arthur Gordon Pym* (1837), Pym compares Augustus's intoxication to that state of madness which "frequently enables the victim to imitate the outward demeanor of one in perfect possession of his senses" (Poe, 1975, p. 50).

[7] We should also remember that the fear-of-eyes theme runs throughout Poe's work of the 1830s and '40s and is not always associated with father-figures. Metzengerstein "turn[s] pale and [shrinks] away from the rapid and searching expression of his [horse's] earnest and human-looking eye" (Poe, 1978, p. 28). The narrator of 'Ligeia' is at first attracted to and then terrified of the black orbs of his first love. And of course, the narrator of 'The Black Cat' impulsively cuts out the searching eye of his pet.

[8] Both Sigmond and the author of 'On Impulsive Insanity' are quoting from an essay by "Dr. Baillarger" which won an award from the French Academy of Medicine for the best dissertation on psychological medicine in 1844 (Sigmond, p. 585). See also Rush's discussion of "uncommonly acute" hearing in *Diseases of the Mind* (1830, p. 143). John E. Reilly (pp. 5—6) has also noticed that the increased acuteness of the senses was thought to be a sign of insanity in Poe's time, but he fails to note that the ticking and, later, ringing sounds heard by Poe's narrator were singled out by Poe's contemporaries as common hallucinations among the insane. He believes the narrator actually hears the noise made by death-watches in the wall, but resorts back to hallucination when he must explain why the ticking increases in tempo just before the murder.

[9] Gunnar Bjurman points out (pp. 220ff) that one source for Poe's plot might have been Daniel Webster's 1830 pamphlet on the trial of John Francis Knapp. Webster describes a self-possessed murderer who, like Poe's narrator, "feels [his crime] beating at his heart, rising into his throat, and demanding disclosure" (XI, 52—54). There is evidence that Poe knew about Webster's pamphlet, but it should be remembered that by 1843, Poe and his audience would have read many such pamphlets and reports. Between 1825 and 1838, the Philadelphia publishing house of Carey and Lea published almost twice as many medical books as those in any other category except fiction, and mental health was a staple concern in these works (Kaser, pp. 72, 119—23).

REFERENCES

'Ancient Case of Homicidal Insanity', *Connecticut Courant*, 15 November 1785, reprinted in *American Journal of Insanity* **3** (1847) pp. 283—4.
'Baron Rolfe's Charge to the Jury, in the case of the Boy Allnutt, who was tried at the Central Criminal Court, for the Murder of his Grandfather, on the 15th Dec., 1847', *Journal of Psychological Medicine and Mental Pathology* **1** (1848) pp. 193—216.
Bjurman, G.: *Edgar Allan Poe: En Litteraturhistorisk Studie*, Gleerup, Lund, 1916.
Carlson, E.: Introduction, B. Rush, *Two Essays on the Mind*, Brunner/Mazel, New York, 1972, pp. v—xii.
Coventry, C.: 'Medical Jurisprudence of Insanity', *American Journal of Insanity* **1** (1844) pp. 134—44.
Fosgate, B.: 'Case of William Freeman, the Murderer of the Van Nest Family', *American Journal of the Medical Sciences* **28** (1847) pp. 409—14.

medical sources Poe drew from are now outdated, and we no longer recognize Poe's medical allusions. But the deep-seated and not always clearly verbalized anxiety generated by the knowledge that men like Prichard and Rush imparted is still with us. What Poe's 1843 audience had learned — what his present audience is still struggling with — was that a murderous rage could be present in any man, could begin to manifest itself without motivation, and once manifest, could exert complete control. The will to do wrong was internally derived; it could no longer be referred to poisonous miasmatas, solipsism, alcohol, or intellectual indulgence. Even reason could provide no check on these murderous rages, since the most careful plans and meticulous arguments could be made to support the most vicious actions. This was, and is, the real terror of Poe's tale: that there is in man the potential for an inexplicable moral short-circuit that makes it impossible to find protection from the dangers that lay within our neighbors — and ourselves. It is to Poe's credit as an artist that he has given this terror an imaginative representation which has remained valid long after Prichard's theories have disappeared.

NOTES

[1] Rush originally published these essays in 1786. They were reprinted in *Medical Inquiries and Observations upon the Diseases of the Mind* in 1812, where they went through five editions and numerous translations. In 1972, Brunner/Mazel reprinted them again as a separate volume, introduced by E. T. Carlson, entitled *Two Essays on the Mind*.

[2] When combined with the notion that each faculty was connected to a particular area of the brain, Rush's theory gained widespread acceptance as phrenology. Poe was at one time an adherent to some of the ideas espoused by phrenology, but by the 1840s, his views were closer to the views of established medicine.

[3] A case fitting this description actually exists. See 'John Ball's Case' (pp. 85—6). See also 'Ancient Case of Homicidal Insanity' (pp. 283—4), which gives the case of a man convicted for murdering his wife despite the fact that he felt she was one of the witches and wizards haunting him.

[4] Thanks to the work of Pinel and the moral managers, public opinion regarding insanity was becoming more enlightened, and as public awareness increased, defense pleas of insanity became more common. There were only a few such cases before 1825, but by the late 1840s there were well over fifty.

[5] For a good example of how newspapers reported on these trials, see the reports of the Freeman trial in the [*New York*] *Evening Post*, 19 March 1846, p. 1, col. 9, and the *New York Tribune*, 20 March 1846, p. 3, col. 1.

[6] It is clear, however, that Poe knew something about moral insanity as early as 1837.

Wharton, F.: *A Treatise on Mental Unsoundness Embracing a General View of Psychological Law*, 2 vols., Kay & Brother, Philadelphia, 1873.
Woodward, S.: 'Moral Insanity', *Boston Medical and Surgical Journal* **30** (1844) pp. 323—36.

The University of North Carolina at Chapel Hill

Haslam, J.: 'The Nature of Madness', in *Madness and Morals: Ideas on Insanity in the Nineteenth Century* (ed. by V. Skultans), Routledge and Kegan Paul, Boston, 1975, p. 31. (Excerpted from J. Haslam, *Observations on Madness and Melancholy*, Callow, London, 1810.)
'Homicidal Insanity, Case of Hadfield', *American Journal of Insanity* 3 (1847) pp. 277—82.
'John Ball's Case', *New York City-Hall Recorder* 2 (1817) pp. 85—6.
Kaser, D.: *Messrs. Carey & Lea of Philadelphia: A Study in the History of the Booktrade*, Univ. of Pennsylvania Press, Philadelphia, 1957.
'On Impulsive Insanity', *Journal of Psychological Medicine and Mental Pathology* 1 (1848) pp. 609—22.
Poe, E.: 'Metzengerstein', in *The Collected Works of Edgar Allan Poe* (ed. by T. O. Mabbott), Harvard Univ. Press, Cambridge, 1978, Vol. 2, pp. 15—31.
Poe, E.: *The Narrative of Arthur Gordon Pym* (ed. by H. Beaver), Penguin, Baltimore, 1975.
Poe, E.: 'The Tell-Tale Heart', in *The Collected Works of Edgar Allan Poe* (ed. by T. O. Mabbott), Harvard Univ. Press, Cambridge, 1978, Vol. 3, pp. 789—99.
Prichard, J.: *A Treatise on Insanity and Other Disorders Affecting the Mind*, Sherwood, Gilbert, and Piper, London, 1835.
Reid, J.: *Essays on Hypochondriasis and Other Nervous Affections*, Longman, Hurst, Rees, Orme, and Brown, London, 1823.
Reilly, J.: 'The Lesser Death-Watch and "The Tell-Tale Heart"', *American Transcendental Quarterly* 2 (1969) pp. 3—9.
Robinson, A.: 'Poe's "The Tell-Tale Heart"', in *Twentieth Century Interpretations of Poe's Tales* (ed. by W. Howarth), Prentice-Hall, Englewood Cliffs, 1971, pp. 94—102.
Rush, B.: 'An Enquiry into The Influence of Physical Causes upon the Moral Faculty', in *Two Essays on the Mind*, Brunner/Mazel, New York, 1972, pp. 1—40.
Rush, B.: *Medical Inquiries and Observations upon Diseases of the Mind*, 4th edn., John Grigg, Philadelphia, 1830.
Saliba, D.: *A Psychology of Fear: The Nightmare Formula of Edgar Allan Poe*, Univ. Press of America, Lanham, 1980.
Shulman, R.: 'Poe and the Powers of the Mind', *ELH* 37 (1970) pp. 245—62.
Sigmond, G.: 'On Hallucinations', *Journal of Psychological Medicine and Mental Pathology* 1 (1848) pp. 585—608.
'The Trial of James Wood', *Proceedings of the American Antiquarian Society* 52 (1843) pp. 105—6.
The Trial of William Freeman, for the Murder of John G. Van Nest, including the Evidence and the Arguments of Counsel, with the Decision of the Supreme Court Granting a New Trial, and an Account of the Death of the Prisoner, and of the Post-Mortem Examination of His Body by Amariah Brigham, M.D., and Others (reported by B. Hall), Derby, Miller & Co., Auburn, 1848.
Watson, A.: 'Three Medico-legal Cases of Homicide, in which Insanity was pleaded in Exculpation', *Edinburgh Medical and Surgical Journal* 38 (1832) pp. 45—58.
Webster, D.: *Writings and Speeches*, National Edition, 18 vols., Little Brown and Co., Boston, 1903.

resist wider interpretations. Beginning in the early part of this century, however, American writers have alternately approached and retreated from a recognition that technology is not simply tool-making, nor even complicated machinery, but something that transforms all aspects of culture, including a culture's literature. Literary responses to changes in conception of technology have grown increasingly complex — but slowly.

The standard eight-volume *History of Technology* (Singer) points out that technology came to be known as 'applied science' only during the nineteenth century. Indeed, to judge from the passage from the U.S. Constitution (written 1787) cited above, it would appear that the founders of the Republic equated science with literature, both to be distinguished from the "useful Arts". Thomas Jefferson and other scientists, responding to an anti-intellectualism already well established in America, felt constrained to justify their activities in terms of practicality, to insist that discoveries inside a laboratory had 'useful' applications outside. In a real sense, early American science was little more than technology, as numerous historians have established (Greene; Hindle; Reingold; Struik), but popular pressures for democratic access encouraged scientists to cede some territory to inventors while increasing the professional distance between themselves and the latter. Prior to 1800, then, and for many decades afterward, technological pursuits were explicitly open to the amateur, who might be educated, but just as often was not. Inventors were 'mechanics' and technology was tinkering, something more than that if the result were useful or elegant, something less if it were not. Before the nineteenth century, mechanics still shared knowledge, and transmitted it through apprenticeships. But the explosive growth of information began to destroy the concept of shared knowledge; the Constitution's provision for patent and copyright is a recognition of the proprietary nature of new information. Jefferson, as the nation's first Secretary of State, was also her first patent commissioner; despite those duties, he resisted the granting of patents because intellectual property rights seemed to threaten democratic access to knowledge. In founding the University of Virginia, Jefferson tried to offset those threats by spreading ideas through wide education.

The writer defined his function in a culture altering its basis of communal knowledge by insisting not only on his romantic individuality, but also on his originality. Originality was what he marketed

JOSEPH W. SLADE

CONCEPTUALIZING TECHNOLOGY IN LITERARY TERMS: SOME AMERICAN EXAMPLES

How a writer conceptualizes technology, perhaps even more than how he thinks of his own pursuits, will determine whether he perceives or denies affinities between technology and literature. To some degree, such conceptions are functions of larger national concerns and social attitudes. For the framers of the Constitution of the United States, writers and inventors enjoyed certain privileges not accorded to other citizens. Theirs are the *only* two professions singled out by that document as entitled to special protection. Because books and inventions contribute to the public good, their creators are entitled to profit through copyright and patent; Article I, Section 8, Paragraph 8 gives Congress the power "To promote the progress of Science and useful Arts, by securing for limited Times to Authors and Inventors the exclusive Right to their respective Writings and Discoveries". Their common property rights notwithstanding, American writers traditionally have not only declined to affirm the relationship, but have also viewed their colleagues with suspicion. The explanation is largely historical; for many years technology held no fixed place in American cultural hierarchies. That uncertainty, in turn, grew out of even more basic confusion about what, exactly, technology was. While the definition of imaginative literature as a symbol-making activity has remained reasonably constant over the last hundred years, conceptions of technology have been revised (beginning at least with Ernst Kapp's *Outlines of a Philosophy of Technology* in 1877) to include activities and innovations beyond simple tool-making. Since poets and novelists and literary critics seldom read scholarly treatments of technology, they have tended to rely on popular notions, which rarely agree with professional assessments.

To say so is not to accuse writers of excessive snobbery, especially when critics and historians of technology continue to argue over the differences between technology and science. Still, even when they are not particularly hostile, writers have seemed prone to view technology according to fairly narrow cultural and esthetic assumptions, and to

Frederick Amrine (ed.) Literature and Science as Modes of Expression, 153–168.
© 1989 *Kluwer Academic Publishers, all rights reserved.*

Nineteenth-century writers, especially those drawn by the local color movement, might embrace regional speech and homely customs, but most made little attempt to understand the vernacular of technology. The major exception among writers of the previous century was Walt Whitman, whose artistic gaze was unhampered by considerations of social or economic class. His poetry celebrates the mechanical along with the organic, the artificial along with the natural. More typical was Ralph Waldo Emerson, whose transcendentalism could on occasion conflate machinery with spirit in prophecies that Leo Marx has called "the technological sublime", but whose more sober appraisals concluded that technology dulled the senses. When he observed in his 'Ode Inscribed to W. H. Channing' that "Things are in the saddle,/ and ride mankind", he was giving voice to the transcendentalist's mistrust of materialism, but he betrayed as well an elitism regarding the artifacts of low culture. Ambivalence like Emerson's doubtless derived in part from concern about the writer's own role as symbol-maker, especially when technologies appeared to poach on that territory. Airplanes, elevator buildings, and home appliances followed the locomotives that Emerson found so metaphorical, and works of fiction and poetry could not ignore their symbolic importance. Not surprisingly, given the romanticism of the nineteenth century, Emerson and his successors focused on the machine's impact on the individual consciousness; they were less adept at calculating the effects of technology on that domain of collective meaning which is culture.

As machinery grew more complicated and ubiquitous, however, and began to transform the institutions of American society on a large scale, the need to evaluate cultural consequences became more urgent for writers. Here, just as the century turned, popular and literary views converged. The public could still subscribe to the cult of the heroic inventor, finding in the Singers, Edisons, and Fords examples of creativity, but remaining unaware of the data bases upon which their inventions were built. The writer might or might not recognize the creativity so easily obscured by the tedium and the misery associated with factories, but he too generally failed to understand that assembly lines embodied knowledge. To confuse presumed effects with technology itself is common enough even today. In this respect, technology is a term like pornography — almost everyone conceives of it differently. Where some see technology as saving or extending or even ennobling human labor, others see it as dehumanizing. Where some see it as an

(Graña). The professionalization of both science and literature, begun during the eighteenth century, enabled scientists and writers gradually to see themselves as members of an aristocracy of the intellect.[1] To professionalize in both cases was not merely to assert intellectual superiority but also to see oneself as heir to a systematized body of knowledge or experience that was closed to outsiders.[2] One could in fact make a case, despite the celebrated schism between the two cultures, that writers have at least since the nineteenth century felt more of cultural and esthetic kinship with scientists than with inventors, and that both groups, as they learned to ground their disciplines on critical and theoretical principles, increasingly looked down on grubbier mechanical skills.[3] The inventor was not an intellectual in popular estimation, not when compared to the writer. Like the scientist, the writer frequently enjoyed high status, often as compensation for low income, while the mechanic's social standing was blue collar. If writers came to take pride in their elevated tastes, scientists could emphasize the 'purity' of their disciplines; both could glory on occasion in their *not* being immediately useful.

That the design and instrumentation of experiments is crucial to the method that makes science possible, and that language is the technology of expression that makes literature possible, are now becoming obvious. In the nineteenth century, however, the writer, like the scientist, was part of a professional elite that could ignore such mundane considerations. To be a writer was to be a conduit of the muse, to be inspired, to be gripped by imagination, to improve the sensibilities of one's fellows. By contrast, as Robert Multhauf has noted, there was not even a word to designate the role of the person who "improved technology". Americans understood that writers and scientists made contributions to experience and to knowledge (in large part because writers and scientists cultivated such images), but until the end of the nineteenth century they probably would not have agreed that a mechanic or an inventor or a craftsman or artisan added to intellectual endeavor, because those titles did not imply organized effort extending over time, let alone the gradual accumulation of knowledge and expertise. The opposite view prevailed: invention was best when it was spontaneous and practical — and without much intellectual significance. John Kouwenhoven has suggested that American mechanics throughout the eighteenth and nineteenth centuries were working in a kind of vernacular, a demotic art form, whose hallmark was its utilitarianism.

steam engine became lethal in the form of the machine gun, and the age of steam reached a kind of symbolic end (Slade). That shift, while decisive, did not preclude the occasional nostalgia for earlier technologies like the plow, which surfaced in *I'll Take My Stand* (1930), whose twelve southern authors seemed determined to rekindle agrarianism by their very rage against modernity. (Older technologies, because they are simpler and more familiar, generate nostalgia and retain popular appeal — witness Woody Allen's recent film, *Radio Days* (1985), a throwback to the halcyon days before television.)

The nationalism associated with technology, detailed by John F. Kasson, continued to manifest itself, but in more personal and individual terms. After the turn of the century, technological aptitude finally became professionalized in the figure of the engineer. In fact, the engineer achieved cult status as the consequence of efforts by men as diverse as Frederick W. Taylor, Henry Gantt, Howard Scott, and Thorstein Veblen. Taylor, through his time-and-motion studies, tried to introduce efficiency into various corners of culture. Movements like Gantt's ("The New Machine") and Scott's ("Technocracy"), which would eventually lead to the election of Herbert Hoover (an engineer) as President of the United States, exalted the engineer as an improver of technology, a contributor to knowledge, and a designer of culture. Such qualities, advocates like Scott insisted, equipped the engineer to function as a social and political statesman. Veblen and his followers deflected the fear of rampant technology by transferring the blame for side effects onto American commercialism. The engineer, said Veblen, in works like *The Theory of Business Enterprise* (1904) and *The Engineers and the Price System* (1921), was a creative genius, as driven by esthetic impulse and a desire to improve culture as any artist. It was the parasitical American businessman, who, in his desire to exploit new inventions for profit, perverted the morally neutral technology into grotesque forms. The engineer was a hero, his inventions and designs worthy of emulation.

The engineer's rationality and efficiency seemed morally and esthetically admirable after world war, political uncertainty, and economic adversity. During the 'twenties, this image of *homo faber* at last became appealing to writers. The whole culture learned to admire the engineer's practicality, and to accept the notion of efficiency as a way of overcoming the messiness of human society. In her recent book, Cecelia Tichi has examined the powerful influence of what might be called a mechan-

agent of civilization, others see it as a despoiler of nature. Few comprehend that technology, whether liberating or oppressive, *is* knowledge.[4] Literature seems to occupy a realm of ideas, but technology seems distressingly physical, best treated — for writers, at least — as metaphor.

The transformation of technology into a body of knowledge whose size and consistency resembled those integral to science and literature was a gradual one. By 1900, according to Edwin Layton, engineering "constituted a complex system of knowledge, ranging from highly systematic sciences to collections of 'how to do it' rules in engineering handbooks".[5] That was more apparent perhaps to engineers than to inventors, who to this day like to claim for their products a novelty without precedent. It was certainly not an idea widely accepted by the public or by writers, who continued to divide the world into the impractical and the utilitarian. At the beginning of the twentieth century, however, even the most practical technologies could undermine intellectual assumptions, could, in fact, subvert established traditions, and sometimes did so with stunning effect, as works like *The Education of Henry Adams* demonstrated.

Cultural historians and literary critics, bedazzled by his images of the virgin and the dynamo, and befuddled by his musings on entropy, have too often characterized Henry Adams as a twentieth-century visionary instead of the nineteenth-century Tory he was. Where Adams saw chaos, most historians see only mild discontinuities, the normal skewing caused by advancing technology. For all the anguish that the image of the dynamo and the prospect of entropy caused Adams, similar fears about technology and science voiced by his literary forebears were being modified by history. The nineteenth-century characterization of the machine as the invader of the garden, so ably chronicled by Leo Marx, was fading, eroded by geographical and social factors. The western frontiers had closed, and the agricultural population of America was moving into the cities, a process that has continued almost uninterrupted to the present, when seventy five per cent of the citizenry of the United States lives in urban areas, and less than four per cent make their living as farmers (Porat). Americans quite consciously chose the politics of urban industrialism over rural traditions in the national elections of 1896 and 1900. From 1912 on, writers themselves mounted a revolt from the village, drawn unwillingly perhaps into the wider arenas opened by World War I, which industrialized death. The

is ironic. Less negative is Lewis's engineer novel, *Arrowsmith* (1925), or a later work like Agee's *Death in the Family* (1956). Inventions were assimilated by the culture and by literature quickly — telephones being a prime example in that last novel.

In another recent critical work — one of the amazing things about the field of literature and technology is the speed with which critics are now mining it — Lisa Steinman focuses on the "machine aesthetic" of William Carlos Williams, Marianne Moore, and to a lesser extent, Wallace Stevens. To such poets, a well-crafted poem possessed the moral symmetries and beautiful lines of a well-wrought machine, and could be justified in the same practical terms. Poems, said Williams, were "machines made of words". One of the principal motivations behind such statements was the need to break away from nineteenth-century aesthetic principles, and to widen the arena of modern literary activity. Steinman is nevertheless convinced that in emulating engineers and stressing affinities between literature and invention, American writers were attempting to appropriate the growing authority attributed to technology. The attempt embraced a telling contradiction, however, in that it also represented envy of the engineer's commercial success. Writers wanted literature to be valued for its practical utility, and to command better prices.

The commercial value of literature, or the seeming lack of it, was very much at issue, and it rendered the position of the writer in America far more complicated than most literary historians have indicated. Neither Tichi nor Steinman recognize that it was not simply the humanizing of the mechanical in the figure of the engineer, nor the social utility nor the commercial value of his tools that made technology suddenly respectable: it was also the realization that technological creativity rested on knowledge, not just apprenticeship or crude experience. An invention embodied information. Thinking about products in this way makes it easier to comprehend the shifting structures of the American commodity system, as industrialization began to give way to post-industrialization, a term which refers to the marketing of information itself.

Behind the individual invention was the larger question of industrialization. Here Veblen's legacy was powerful, augmented by the writer's inclination to attribute evil to businessmen. But in fact industrialization was entering a new phase whose outlines seem obvious only in retrospect. Writers were bewildered by its onset, which was magnified by the

ical muse on writers of imaginative literature during the 'twenties. Tichi suggests that this influence went beyond the adoption of mechanical metaphors by novelists and poets. In her view, writers began to construe literary works as "assemblies of component parts, including prefabricated parts",[6] and the writer as a craftsman who welded tropes together. The most famous manifesto was that announced by Hart Crane, the self-styled "Pindar of the Machine Age", who wrote in his essay "Modern Poetry":

> For unless poetry can absorb the machine, i.e., *acclimatize* it as naturally and casually as trees, cattle, galleons, castles and all other human associations of the past, then poetry has failed of its full contemporary function. This process does not infer any program of lyrical pandering to the taste of those obsessed by the importance of machinery; nor does it essentially involve even the specific mention of a single mechanical contrivance. It demands, however, along with the traditional qualifications of the poet, an extraordinary capacity for surrender, at least temporarily, to the sensations of urban life. This presupposes, of course, that the poet possesses sufficient spontaneity and gusto to convert this experience into positive terms. Machinery will tend to lose its sensational glamor and appear in its true subsidiary order in human life as use and continual poetic allusion subdue its novelty. For, contrary to general prejudice, the wonderment experienced in watching nose dives [by airplanes] is of less immediate creative promise to poetry than the familiar gesture of a motorist in the modest act of shifting gears [the source of Tichi's title]. I mean to say that mere romantic speculation on the power and beauty of machinery keeps it at a continual remove; it can not act creatively in our lives until, like the unconscious nervous responses of our bodies, its connotations emanate from within — forming as spontaneous a terminology of poetic reference as the bucolic world of pasture, plow, and barn.[7]

Crane's great poem, *The Bridge* (1930), is evidence of his ability to find meaning — esthetic, intellectual, emotional, and spiritual — in the technology of his time. Crane found beauty in acetylene torches, subways, elevators, neon signs, forges, locomotives, and radios. Behind this monumental achievement, however, lie better examples of the domestication of technology in literature. A novel like Sinclair Lewis's *Babbitt* (1922), for example, is littered with gadgets, none of them possessed of particular significance, but all of them increasingly essential elements in the lives of Americans. Babbitt himself enjoys comparing himself to a "shuttle of polished steel darting in a vast machine," although it is clear enough from the book's many machine metaphors that Lewis intends them as a condemnation of a sterility he associated with a mechanized culture. Babbitt loves inventions with a mindless joy that reflects his own failures of imagination; his kinship with machines

writers during the 'thirties more directly than ever before precisely because they were designed to sell the best literature along with the worst. Technology thus subverted the traditional hierarchies of high and low culture by selling the artifacts of both. The inroads were the consequence of the rapid development of communications media: radio, cinema, publishing. Mass-oriented media were based on technologies whose expense called for corporate finance and management. If the advent of sound boosted demand for motion pictures, the price made AT&T, RCA, and the Morgan banks the real arbiters of taste. Although opera and drama now came to the airwaves by courtesy of Texaco and General Electric, the crucial factor in selling culture was not so much the cost of the medium as the size of its appetite. Programming even a 'classical' radio station like New York's WQXR required 7,000 hours of music a year. The typical Hollywood studio, faced with an annual quota of fifty or sixty features, was desperate to hire a Hemingway, a Fitzgerald, or a Faulkner, and hired they were. New techniques of market research made clear to the publishers of *The New Yorker* that the nation's college graduates (whose number doubled between 1920 and 1930) and professionals (whose number quadrupled) wanted accurate information and quality entertainment — both in large volume.[8] Upset and confused by the new commerce, conservative critics charged that technology undermined ideas, cheapened art, diminished the artist, and prostituted our heritage. The new technologies provided plenty for everyone to complain about: more liberal observers noted that mass distribution stripped serious art of its scarcity and thereby democratized it, but were not so happy that the collaborative nature of the new media altered the role of the individual craftsman and thus deromanticized it.

But hardly any of the critics noticed that traditional literature had also became a commodity sold by merchandizing strategies. Here begins the acceleration of that inflationary spiral which has led to 52,000 new book titles published in the United States in 1985. From its inception in 1926, the Book-of-the-Month Club grew so fast that by 1943 wartime paper shortages forced it to limit membership to 600,000. By then, the BOMC was the third largest patron of the Post Office, surpassed only by Sears Roebuck and Montgomery Ward. The BOMC made George Santayana (among other authors) a best-selling writer. And the BOMC had competitors: the Literary Guild, the Doubleday One Dollar Book Club, and the Book League of America.[9]

Depression, whose seismic effects altered both technology and American economics. The 'thirties are crucial because American business began to industrialize the writer's product and to mass produce and distribute it.

Industrialization, which had of course begun in America much earlier, is not just a system of production and distribution but also one of capitalization. It is as much an economic process as a technological one. Alan Trachtenberg has noted the spread of corporate techniques during the nineteenth and early twentieth centuries, but better sources are Thomas Hughes's *The Vital Few* (1986) or any of the works of Peter Drucker. A great many entrepreneurs — not just the Edisons and Fords, but the Swifts, Eastmans, Carnegies, and Morgans — were inventing the corporation, which is itself a technology of enormous sophistication. The American corporation gradually evolved command and control technologies to increase production during the industrial era, and wedded them to research and development strategies. During the 'thirties, however, the processing of knowledge began to emerge as the crucial function of American business. As the economist Joseph Schumpeter put it about this time, information was becoming the true capital of advanced economic systems (Schumpeter, 1947, 1950). Marxist critiques of industrialism, although favored by many writers, were inadequate to deal with the evolving nature of modern capitalist systems, chiefly because they converted information into ideology. More specifically, Marxist analyses were based on nineteenth-century positivist conceptions of reality, and assumed that science and technology were mere superstructures for materialistic historical forces. From such analyses came condemnations of industrialism as alienating and dehumanizing, expressed in the cliche that Americans themselves were cogs in the machine of culture. Marxist-derived or (just as often) not, assembly-line metaphors had by now become staples of American fiction. While these charges had some truth, they did not grasp very basic changes in the nature of corporate systems. Marxists regarded communications media as just another form of exploitation, a way of feeding pablum to the masses, of reinforcing the status quo. Even critics like Walter Lippmann, who in 1922 denounced what he called the "pseudo-environment" of mediated messages (Harris), appeared to believe that some sort of reality existed apart from what he saw as a false culture.

Corporate technologies impacted on the lives and fortunes of American

Dos Passos's work lies not in its content but in his technique. Numerous readers have commented on his use of the "Camera Eye" and "Newsreel" innovations, but what made those headlines, advertisements, images, tag ends of conversation, radio shows, and so on so appropriate was the implicit suggestion that culture was to be understood as mediated messages, that information shaped the culture, that it was both capital and commodity, a conviction emphasized by the prominence given to the character of J. Ward Moorehouse, the powerful public relations mogul. It was not just that newspapers rewrote American culture every morning; the business of America had become information. Today, of course, only twenty per cent of the American work force is devoted to industrial production (Porat). By far the majority processes information — by means of technology, and *as* a technology; that fact, more than any other, defines American culture.

Following World War II, when the primacy of technology was shadowed by atomic clouds, American culture lurched steadily into the information age. In the late 'forties, Claude Shannon converted the formula for entropy that had so frightened Henry Adams into the very measure of information (Shannon and Weaver). Bell Laboratories invented the transistor, basis of rapid electronic communication systems, and the University of Pennsylvania invented ENIAC, the first viable electronic computer. Eisenhower announced the arrival of the military-industrial complex, and writers began to see that Jefferson's universities were part of post-industrial America, i.e., essential to an information economy, essentially corporations with different strategies and markets. Academic novels like John Barth's *Giles Goat-Boy* (1966) pictured American culture and American technology as interdependent; united, shaped, and determined by structures of knowledge and expertise, by the flow of information along ever larger networks. Data in various configurations, processed as an institutionalized activity, are the principal elements of that novel. The computer, says a 1981 report by the Sloan Foundation, "like writing ... is a technology of thought" (Koerner, p. 4). At the very least, the advent of the computer reminds us that the manipulation of symbols is not so distinct from the manipulation of tools.

That American writers have now become aware of technology as essentially an information-processing activity, and that writing is itself a technology, are ideas most clearly illustrated by the work of Thomas Pynchon, whose three novels explore a culture transformed by informa-

Still more important was the paperback book, pioneered by Pocket Books in 1939, which used magazine presses to mass produce Pearl Buck's *The Good Earth* and other inexpensive editions of classic literature (the sleazy titles actually came later, in the 'fifties). By 1941, Pocket Books was selling ten million copies a year; by 1942, twenty million. And it too had competitors, who sold volumes in such quantity that censorship groups sprang up around America to complain that children could buy dirty books by Ernest Hemingway for only twenty-five cents (Cowley). The reciprocal nature of technologies of information processing and distribution and the culture which both generated and consumed that information became visible.

To a degree, and from a somewhat different esthetic perspective, the New Criticism of the 'twenties, 'thirties, and 'forties was an acknowledgment of the industrialization of literature, a recognition that product could be divorced from worker, that the process of writing was itself somewhat mechanical, that a text could be considered apart from the author. These ideas were driven home by the necessity of translating drama into radio plays, or novels into film scripts; there was something mechanical about the addition and subtraction in adapting literature from one medium to the other. And while these developments should be placed against the burgeoning of a large market for fiction — so large that a critic would attack the mass production of literature in a book called *The Dance of the Machines: The American Short Story and the Industrial Age* (O'Brien) — it was clear that radio and the movies had supplanted traditional literature as the principal story-telling media of America (just as television has supplanted them today). Thanks to structuralism and other academically popular schools of thought, we are more accustomed to the concept of a text as a commodity today, but at the time such a notion would have affronted the writer's sensibilities.

The changed nature of American industrialization and its consequences for literature were grasped imperfectly. John Dos Passos is a pivotal figure, and his *U.S.A.* trilogy (1930, 1932, 1936) a pivotal text. On the one hand, Dos Passos includes with his portraits of statesmen, labor leaders, and visionaries those of inventors and industrialists like Taylor, Steinmetz, and Edison as shapers and fabricators of American culture. On the other, he analyzes American industrialism in conventional leftist terms as exploitative and dehumanizing, although it is clear that he was drifting away from the Marxism of his youth. The novelty of

creates embodies information. American writers are beginning to realize that understanding culture requires understanding technology. But that in turn requires not only acknowledging that technology is knowledge, but that literature is a technology too. As they redefine their own profession, writers may discover their kinship with other inventors.

NOTES

[1] For the professionalization of the writer's trade, see Sontag (p. 22) and Graña (pp. 40—42); for the professionalization of science, see Morton (pp. 1—17) and Kevles (1978).
[2] See Kuhn (pp. 20—22), who suggests that a mature science is no longer accessible to amateurs.
[3] Once disciplines like classics and physics become rigidified, as in the universities of England, they sometimes appear to be partners in a conspiracy to deny primacy to technology — witness Britain's diminished status in the post-industrial era, a great irony, considering her earlier history of industrialization.
[4] See Layton's seminal essay on technology as knowledge (1974).
[5] In his essay on nineteenth-century science and technology, Layton (1971, p. 567) provides other examples of the engineer's professionalized organization of knowledge.
[6] Tichi (p. 267); this profusely illustrated text surveys images of technology across the culture.
[7] Crane (pp. 261—62); see also his 'General Aims and Theories' (pp. 217—23).
[8] The best single account of the commercialization of American high culture is Marquis, although she does not deal to any great extent with literature.
[9] Spiller (p. 1267); this history also contains interesting observations about the 'institutionalization' and 'collectivization' of literary activities (pp. 1270—72).

REFERENCES

Cowley, M.: 'Cheap Books for the Millions', in *The Literary Situation*, Viking Press, New York, 1958, pp. 96—114.
Crane, H.: 'Modern Poetry', in *The Complete Poems and Selected Letters and Prose of Hart Crane*, Doubleday Anchor, Garden City, N.Y., 1966, pp. 260—63.
Drucker, P.: *The Age of Discontinuity: Guidelines to Our Changing Society*, Harper & Row, New York, 1969.
Drucker, P.: *Concept of the Corporation*, John Day Co., New York, 1942.
Drucker, P.: *Technology, Management and Society (Selected Essays)*, Harper & Row, New York, 1970.
Eliot, T.: 'Choruses from "The Rock"', *The Complete Poems and Plays 1909—1950*, Harcourt, Brace & World, New York, 1962, pp. 96—114.
Graña, C.: *Modernity and Its Discontents*, Free Press, Glencoe, Ill., 1967.
Greene, J.: *American Science in the Age of Jefferson*, Iowa State Univ. Press, Ames, Iowa, 1984.

tion as the energy of all of man's — and perhaps God's — systems. In our appreciation of his perspective, however, we should not overlook the degree to which he relies on the insight of T. S. Eliot. As numerous critics have noticed, the landscape of Pynchon's fiction resembles *The Wasteland* of 1920. Just as much a Tory as Henry Adams, but possessed of the expertise of a banker rather than a historian, Eliot — almost alone among writers before WW II — understood that technology meant information, not just dynamos. His wasteland is afflicted by the absence of love and of water, but also by the presence of information — in deluge. Eliot puts it well in "The Rock":

> Endless invention, endless experiment,
> Brings knowledge of motion, but not of stillness;
> Knowledge of speech, but not of silence;
> Knowledge of words, and ignorance of the Word.
> All our knowledge brings us nearer to our ignorance,
> All our ignorance brings us nearer to death,
> But nearness to death no nearer to God.
> Where is the life we have lost in living?
> Where is the wisdom we have lost in knowledge?
> Where is the knowledge we have lost in information?
>   (Eliot, 1952, p. 96).

Echoes of Eliot's complaint about information overload are now commonplace. We might mention just three disparate novels. In one, Ted Mooney's *Easy Travel to Other Planets* (1981), characters come down with 'information sickness', a condition induced by over-exposure to data. In the second, William Gaddis's *J.R.* (1975), Americans are obsessed with information, which is what they buy and sell, as fact or fiction. In the third, Don DeLillo's *White Noise* (1983), messages are ubiquitous; the problem is discovering meaning in them. Culture, of course, is the domain of meaning, but meaning is the product not just of efforts traditionally thought intellectual but also of the most vulgar technologies. Culture itself is a fabrication, the product of artifice that is only occasionally literary. Culture centers on the marketplace, where ideas are exchanged, ideas that are now more than ever commodities shaped and distributed by technological means. To the degree that a society preserves its memories, it stores that culture as a technology, as information (as fact or fiction, as experience or interpretation, as literature, art, science or whatever) in knowledge bases. Everything man

Steinman, L.: *Made in America: Science, Technology, and American Modernist Poets* Yale Univ. Press, New Haven, 1987.
Struik, D.: *Yankee Science in the Making*, Little, Brown, Boston, 1948.
Tichi, C.: *Shifting Gears: Technology, Literature, Culture in Modernist America*, Univ. of North Carolina Press, Chapel Hill, N.C., 1987.
Trachtenberg, A.: *The Incorporation of America: Culture and Society in the Gilded Age*, Hill & Wang, New York, 1982.
Veblen, T.: *The Engineers and the Price System*, Scribner's, New York, 1921.
Veblen, T.: *The Theory of Business Enterprise*, Scribner's, New York, 1904.

Long Island University

Harris, N.: 'From Sermons to Sonys: How We Keep in Touch', *Time* (16 February 1976) pp. 69—71.
Hindle, B.: *The Pursuit of Science in Revolutionary America, 1735—1789*, Univ. of North Carolina Press, Chapel Hill, N. C., 1956.
Hughes, T. *The Vital Few: The Entrepreneur and American Economic Progress*, expand. edn., Oxford Univ. Press, New York, 1986.
Kasson, J.: *Civilizing the Machine: Technology and Republican Values in America, 1776—1900*, Grossman, New York, 1976.
Kevles, D.: *The Physicists: The History of a Scientific Community in Modern America*, Knopf, New York, 1978.
Koerner, J., ed.: *The New Liberal Arts: An Exchange of Views*, Alfred P. Sloan Foundation, New York, 1981.
Kouwenhoven, J.: *Made in America*, Doubleday, Garden City, N. Y., 1948.
Kuhn, T.: *The Structure of Scientific Revolutions*, 2nd edn., Univ. of Chicago Press, Chicago, 1970.
Layton, E.: 'Mirror-Image Twins: The Communities of Science and Technology in 19th Century America', *Technology and Culture* **12** (1971) pp. 562—80.
Layton, E.: *The Revolt of the Engineers: Engineers, Social Responsibility and the American Engineering Profession*, Case Western Univ. Press, Cleveland, 1971.
Layton, E.: 'Technology as Knowledge', *Technology and Culture* **15** (1974) pp. 31—41.
Marquis, A.: *Hopes and Ashes: The Birth of Modern Times*, Free Press, New York, 1986.
Marx, L.: *The Machine in the Garden: Technology and the Pastoral Ideal in America*, Oxford Univ. Press, New York, 1964.
Morton, P.: *The Vital Science: Biology and the Literary Imagination, 1860—1900*, Allen & Unwin, London, 1984.
Multhauf, R.: 'The Scientist and the "Improver" of Technology', *Technology and Culture* **1** (1959) pp. 38—47.
O'Brien, E.: *The Dance of the Machines: The American Short Story and the Industrial Age*, Macaulay, New York, 1929.
Pool, I.: 'Tracking the Flow of Information', *Science* **221** (1983) pp. 609—13.
Porat, M., et al.: *The Information Economy*, OT Special Publication 77—12, Department of Commerce, Washington, D.C., 1977.
Reingold, N., ed.: *Science in Nineteenth-Century America: A Documentary History*, University of Chicago Press, Chicago, 1964.
Schumpeter, J.: *Capitalism, Socialism, and Democracy*, 3rd edn., Harper, New York, 1950.
Schumpeter, J.: 'The Creative Response in Economic History', *Journal of Economic History* **7** (November 1947) pp. 149—59.
Shannon, C. and Weaver, W.: *The Mathematical Theory of Communication*, Univ. of Illinois Press, Urbana, Ill., 1949.
Singer, C. et al: *History of Technology*, 8 vols., Clarendon Press, Oxford, 1954—84.
Slade, J.: 'The Man Behind the Killing Machine', *American Heritage of Invention & Technology* **2** (Fall 1986) pp. 18—25.
Sontag, S.: 'When Writers Talk Among Themselves', *New York Times* (5 January 1986) pp. 1, 22.
Spiller, R. et al.: *Literary History of the United States*, 3rd edn., Macmillan, New York, 1973.

Kimball sees the machine, as did many others, as man's servant, and as an egalitarian force, making available to the masses what had previously been the luxuries of the aristocracy. But what is most interesting from the present point of view is that Kimball sets machinery against the arts as rivals for our allegiance. Arguing against those who hold that "painting and sculpture, philosophy and poetry, embody somehow grander truths" than the mechanical arts, Kimball sounds like Plato reborn in the Promethean fire of American industry. The arts, for all their grandeur, Kimball claims, are "only of themselves an imitation, or, at most, an idealization, of the outside of things, only a surface expression of ideas. There is an element of falsity runs through them all". Machinery, says Kimball, "is not imitation, but the embodiment, of real forces, laws, and principles, which are made to act". In order for inventions to work, "they must not only seem, but be. A lie in them is absolutely fatal. What would be the worth of a sewing machine, however highly ornamented, which, like a picture, only looked as if it sewed?" (pp. 323—4). To Kimball, in short, machines act upon things, the arts provide merely pictures of things.

One doesn't want to belabor Kimball's ontological confusions, which find their comic *reductio ad absurdum* in Rube Goldberg's improbable machines; but how interesting that Kimball felt some kind of choice and even comparison between the machine and the arts was necessary. (We are reminded, perhaps, that artisan [skilled laborer] and artist [original creator] share a common root.) It is as if the practical and the aesthetic were competing within the culture for influence and ascendancy. The enthusiasm of a Kimball — and he was representative of a widespread attitude — could well cause the writer either to retreat into a cloistered aestheticism (as did many) or else attempt in some other way to come to grips with the changed material conditions of society. And yet, that very effort to absorb the pressures of the modern, of technology, could result in works that are peculiarly, though interestingly, muddled, especially at the end of the nineteenth century. No wonder Mark Twain — who registered acutely so many cultural conflicts — would write a book as conflicted as *A Connecticutt Yankee in King Arthur's Court*, which begins as a vindication of technological rationality in the face of benighted European superstitions and social habit and ends in a nightmarish vision of technology run amok. And other writers too at the end of the century — Dreiser and Norris, for example — would feel the attraction of scientific and technological forces within American

MILES ORVELL

LITERATURE AND THE AUTHORITY OF TECHNOLOGY

Whitman, who is the starting point for modern American literature in so many ways, took it upon himself in *Democratic Vistas* (1871) to define the character of American culture in a way that has had continuing relevance throughout the twentieth century: "America demands a poetry that is bold, modern, and all-surrounding and kosmical, as she is herself. It must in no respect ignore science or the modern, but inspire itself with science and the modern. It must bend its vision toward the future, more than the past". Starting with the close observation of nature, the poet works by analogies, by indirections; his genius lies in the "image-making faculty, coping with material creation, and rivaling, almost triumphing over it" (pp. 503 and 510). Thus, as Whitman describes it, the poet or artist is making a kind of facsimile world, a world that rivals the creation, and indeed — in a curious phrase — almost triumphs over it; inspired by "science or the modern", Whitman posits an analogy between artistic creation and the technological process that situates him at a particular juncture in American culture, the beginning of the modern vision.

For Whitman was writing at a moment when the force of technology was being acutely recognized as a vigorous determinant of social direction, transforming the material fabric of society in starkly visible ways, and displacing even the traditional social power of poetry and the arts; what defines Whitman's response as "modern" is precisely his effort to absorb the meaning of technology — even if only in some vague way — into his aesthetic process, and it is a strategy that I want to track to its culmination in the work of John Dos Passos and James Agee.

We can better understand the pressure behind such a strategy by looking briefly at the post-Civil War period in America, when the revolutionary powers of technology were first being consistently articulated within the culture, and in a way that might well put the arts on the defensive. Consider, for example, the review written by the Reverend John C. Kimball, following his visit to the Eleventh Exhibition of the Charitable Mechanic Association in Boston in 1869.

object, a machine-like object, in a somewhat static sense. The poem has a thing-like objectivity, it is a functional system, an integral design of part and whole in which no part is redundant; but it is essentially a thing to look at, rather than, for example, an instrument for knowing the world.

In order to see how far the authority of technology could carry the writer, we must turn to John Dos Passos and James Agee, who, in their work during the 'thirties, in different ways, sought to develop a mode of writing that established an even deeper relationship to technology, and one that goes back for its inspiration to Walt Whitman.

Dos Passos' clearest formulation of his views in this respect comes in a 1935 speech he delivered at the American Writers' Congress, "The Writer as Technician". For Dos Passos, the professional writer is involved in a process of "discovery and invention", a process "not very different from that of scientific discovery and invention". Here is Dos Passos the modernist speaking, who in fact invented, in the structure of *USA*, the most elaborate novel — machine of its time, a four-part mechanism of fictional narratives, biographies, newsreels, and autobiographical fragments. But it is Dos Passos' view of the writer's relation to society that is particularly interesting in this context: "In his relation to society a professional writer is a technician just as much as an electrical engineer is" (1935, p. 79). The authority that Dos Passos thus claims is borrowed from the larger authority of technology — with its objectivity, its rational solution to problems of disorder and waste, its claims to efficiency — all of which were so compelling in the Depression.

Yet at the same time, Dos Passos wants to preserve the traditional vatic role of the writer as prophet, as seer, a Whitmanesque role that places him in opposition to the domination of machinery. For the successful literary work, according to Dos Passos, will have an influence on "ways of thinking to the point of changing and rebuilding the language, which is the mind of the group". Far from being a mere technician, then, the writer lays claim to a special authority: "At this particular moment in history, when machinery and institutions have so outgrown the ability of the mind to dominate them, we need bold and original thought more than ever. It is the business of writers to supply that thought, and not to make of themselves figureheads in political conflicts" (1935, p. 81). The technician here has become the poet-legislator. At the bottom of Dos Passos' conception of the writer's role is the implication that the literary work can embody a special way of knowing the world.

society, which they would yet try to assimilate within a vocabulary and world-view which was in many ways romantic and mystical.

With the modernist writers of the early twentieth century — beginning with Ezra Pound — technology takes on immediate and profound importance. And Pound's influence on a younger generation of writers who attain maturity in the 'teens and 'twenties — most notably Eliot, Williams, and Moore — has been well documented (see Tichi). But even someone like Willa Cather — whom we think of as a traditionalist, alienated from the modern world of technology and bureaucracy, looking backward to an idealized, organic civilization — even Cather had absorbed by 1925 much of the ethos of the modernists. For in the same work in which she can condemn science as giving us nothing but sleights of hand (*The Professor's House*), she celebrates the virtues of a functional style in a manner that relates her to technological values. As the professor annotates Tom Outland's diary, in which the discovery of the Blue Mesa civilization is detailed, he appreciates a beauty in Outland's minute descriptions of the tools, pottery, and cloth: "If words had cost money, Tom couldn't have used them more sparingly. The adjectives were purely descriptive, relating to form and colour, and were used to present the objects under consideration, not the young explorer's emotions". Yet through the austerity of the style, Cather explains, one feels the discoverer's excitement (p. 262).

Cather's description of Outland's style is a perfect description of Hemingway's style, with its economies of diction and rhythm, its clarity, its sense of emotional pressure under control; and it closely resembles as well the rules that Ezra Pound had formulated over a decade earlier in *Poetry* magazine as "A Few Don'ts": "Use no superfluous word, no adjective which does not reveal something.... Use either no ornament or good ornament.... Consider the way of the scientists rather than the way of an advertising agent for a new soap. The scientist does not expect to be acclaimed as a great scientist until he has *discovered* something" (pp. 106—7).

Still, there is an instructive difference between Pound and Cather. Where Cather associates her functionalist style with the putative organic civilization of the Indian, the younger Pound is more clearly borrowing the prestige and language of science, and his formulation leads to Williams' later notion that "A poem is a small (or large) machine made of words" (p. 256). But this modernist ideal, which so clearly invokes the authority of technology, conceives of the poem as an

immediacy, could avoid that problem. "One reason I so deeply care for the camera is just this. So far as it goes (which is, in its own realm, as absolute anyhow as the traveling distance of words or sound), and handled cleanly and literally in its own terms, as an ice-cold, some ways limited, some ways more capable, eye, it is, like the phonograph record and like scientific instruments and unlike any other leverage of art, incapable of recording anything but absolute, dry truth" (p. 234). Of course, Agee does not assume that the camera is in everyone's hands a recorder of truth: it was Walker Evans, his collaborator, he had in mind, and not, for example, Margaret Bourke-White.

Agee invokes Evans' images — which are printed as part of *Famous Men* — on several occasions, either to supplement his prose descriptions or to replace them, but in fact Agee's prose inevitably has qualities nowhere visible (or possible) in Evans' or any photographer's image. Yet even allowing for the difference in medium, and allowing for the differences in style between Agee and Evans (which I will not go into here), Agee's invocation of the camera forms the basis for his whole moral aesthetic: a deep and self-conscious respect for the actual lives he is depicting. *Famous Men* sums up not only the documentary impulse of the 'thirties, with its urge to encompass the social reality of life in Depression America, but also the modernist urge to invent forms that would reflect the new scientific and technological conditions of knowledge and that would embody, self-consciously, our experience of the world. And here I might add that Agee's constant — and for some readers irritating — intrusion of himself into the exposition is a deliberate strategy, not unlike Dos Passos' use of the "Camera Eye" sections in *USA*. Both books are really as much about their authors as about their subjects; and they both thus make a virtue of Heisenberg's uncertainty principle, which declares that the observer himself is an inevitable function in the calculus of observation, that the eye of the seer makes the thing seen.

Out of the artist's complex sensibility and his attunement to the age would arise, in Agee's words, "the beginning of somewhat new forms, of which the still and moving cameras are the strongest instruments and symbols. It would be an art and a way of seeing existence based, let us say, on an intersection of astronomical physics, geology, biology, and (including psychology) anthropology, known and spoken of not in scientific but in human terms" (p. 245). Whatever else he may have had in mind, Agee was doubtless thinking of the book he was at that moment writing — *Let Us Now Praise Famous Men* — and his words

In *USA* that special knowledge is carried in the "Camera Eye" sections, which provide a personal, subjective, autobiographical perspective on the panorama that otherwise unfolds in the novel. In thus naming these autobiographical units Dos Passos was doubtless attempting to capitalize on the aesthetic authority of photography during the early decades of the century as an instrument for objectively registering the visual world; and probably too he was borrowing the political authority of the moving picture camera that was a part of the influential Soviet documentary cinema movement (Kino-Pravda) anchored by Dziga Vertov. (And incidentally he was reversing Whitman's earlier scorn for the medium, when the latter had said, in *Democratic Vistas*, that the artist's process of creation was "No useless attempt to repeat the material creation, by daguerreotyping the exact likeness by mortal mental means".) Using the camera as symbol, Dos Passos invoked the objectivity of the machine, its capacity to record literally the world before the lens; but he was also invoking, obviously, the process of seeing, which was, he argued, a subjective process.

The key passage is in "Camera Eye 47": "from the upsidedown image on the retina painstakingly out of color shape words remembered light and dark straining" (1946, p. 174). The implied sense of vision itself as something that must be turned right side up, interpreted, was a persistent part of Dos Passos' epistemology: "Your two eyes are an accurate stereoscopic camera, sure enough", he wrote in "Satire as a Way of Seeing" (1937), "but the process by which the upsidedown image on the retina takes effect on the brain entails a certain amount of unconscious selection. What you see depends to a great extent on subjective distortion and elimination which determines the varied impacts on the nervous system of speed of line, emotions of color, touchvalues of form. Seeing is a process of imagination" (1964, pp. 20–21). The artist is thus constructing his vision, his eye formed by the climate of visual stimuli in the culture at large. Combining the objectivity of the camera with the subjectivity of constructed vision, the artist unites the technological and the artistic.

For James Agee, the camera was also a crucial model for writing, and it is explicitly developed as such in his documentary (or antidocumentary) portrait of three Alabama tenant farmers and their families, *Let Us Now Praise Famous Men*. Aiming for an authentic representation of the thing itself, Agee was acutely aware of the limitations of language, which, he held, must inevitably sag beneath the weight of naturalistic description. The visual image, in its wordless

have application as well to the multi-faceted constructions of Dos Passos. And too, Agee's words respond to Whitman's challenge of the previous century, to write a poetry that is "bold, modern, and all-surrounding and kosmical, as America is herself", a poetry that is inspired by science and the modern.

Whether or not they were gaining an audience commensurate with their goals, Dos Passos and Agee were defining in their works a literature that would span the complementary impulses of the technological and the aesthetic. In depicting the artist as an engineer, or as a new kind of scientist, they are not only borrowing the new mystique of science and technology; they are also bringing together in the single person of the writer the components of American culture that were sundered by the John Kimballs, and that Van Wyck Brooks had worried about at the start of the modern period, in *America's Coming of Age* — the technical and the theoretical, the scientific and the artistic, the practical and the spiritual. The work of these modernists thus accepted Whitman's challenge of accurately defining "material creation" in a culture that was coming more and more, through the screen of mass culture, to thrive on caricatures of itself.

BIBLIOGRAPHY

Agee, J. and Evans, W.: *Let Us Now Praise Famous Men*, Houghton Mifflin, New York, 1960.
Cather, W.: *The Professor's House*, Knopf, New York, 1925.
Dos Passos, J.: *The Big Money, USA*, Houghton Mifflin, Boston, 1946.
Dos Passos, J.: 'Satire as a Way of Seeing', in his *Occasions and Protests*, Henry Regnery, Chicago, 1964, pp. 20—32.
Dos Passos, J.: 'The Writer as Technician', in *American Writer's Congress* (ed. by H. Hart), International Publishers, New York, 1935.
Kimball, J.: 'Machinery as a Gospel Worker', *Unitarian Christian Examiner*, November 1869, pp. 323—4.
Pound, E.: 'A Few Don'ts', in *Prose Keys to Modern Poetry* (ed. by K. Shapiro), Harper & Row, New York, 1962, pp. 106—7.
Tichi, C.: *Shifting Gears: Literature, Technology, Culture in Modernist America*, Univ. of North Carolina Press, Chapel Hill, N. C., 1987.
Whitman, W.: 'Democratic Vistas', in *Leaves of Grass and Selected Prose* (ed. by J. Kouwenhoven), Modern Library, New York, 1950.
Williams, W.: 'Author's Introduction' to *The Wedge*, in *Selected Essays*, New Directions, New York, 1969.

Temple University

building blocks of everyday discourse, also inhibit disco
freedom of expression. "Let them snarl at you", the spe
you snarl back at them". The act of snarling is the ac
poem which snarls at its own graphemes. "Crossing a s
light ... is all right", the speaker goes on. "Freedom in
freedom, however, is not without its risks, as we will see.

Spicer's radical suspicions of linguistic boundaries eve
human imagination itself, and its interrelationships with
cosmos which it appears to exist 'in'. Spicer describes a
for the poet-in-the-world:

> ... from what I've seen ... there's no question that objective even
> order for poems to be written. Robin [Blaser] in "The Moth Poer
> coming in the wildest places, something where the odds would be
> one of the moths being just exactly in the place that he wanted the p
> was there a couple of times when it happened. And I think that it i
> that the objective universe can be affected by the poet. I mean —
> made the trees and stones dance, and so forth — and this is son
> almost all primitive cultures, and it, I think, has some definite basis to it

Although Spicer looks back to Greek mythology in sug
poetical imagination and the so-called objective worlc
stitute two separate systems, at least one quantum pl
that such a reciprocal metasystem may indeed be rea
and the Implicate Order, David Bohm theorizes that

> the body enfolds not only the mind but also in some sense the entir
> ... both through the senses and through the fact that the constituen
> are actually structures that are enfolded in principle throughout all spa

Bohm points out that even in commonsensical ever
reciprocity between the crude categories of 'mind' and '
a rare occurrence:

> ... we know it to be a fact that the physical state can affect th
> sciousness in many ways (the simplest case is that we can become c
> excitations as sensations). Vice versa, we know that the content of
> affect the physical state (e.g., from a conscious intention nerves may
> may move, the heartbeat change, along with alterations of gland
> chemistry, etc.) (p. 208).

For both the poet Spicer and the physicist Bohm, then,

STEVEN CARTER

"A PLACE TO STEP FURTHER": JACK SPICER'S QUANTUM POETICS

1

Both the themes and the techniques of Jack Spicer's poetry are dedicated to the "erasures", as Michael Davidson (p. 128) has written, of accepted ideas and structures which are imposed upon language by Newtonian epistemological systems. Spicer's work challenges, as William V. Spanos observes (p. 1), "the metaphysical or logocentric forms that have dominated the poetry — and above all the hermeneutics — of the Western literary tradition". In this paper I want to suggest that Spicer's poetry and poetics challenge the hermeneutics of the Western scientific tradition as well, by presenting the reader with linguistic models which are relevant to a quantum not a Newtonian universe.[1]

Readers of Spicer's poetry occasionally encounter Einsteinian cosmology head-on: "Distance, Einstein said, goes around in circles. This/Is the opposite of a party or a social gathering" (1975, p. 227). Elsewhere, on a grimmer note, Spicer asks his readers to enter "The unstable universe", which "has distance but not much else" (1975, p. 236). It is a lonely universe mirrored in "The tidal swell" of Stinson Beach, which itself is constituted of "Particle and wave/Wave and particle/Distances" (1975, p. 227). The appearance of quanta in Spicer's work is not merely thematic; the chiasmus — the speaker crosses over from particle and wave to wave and particle — mirrors of course the way electrons "cross over" from one manifestation, a wave, to another, a particle, depending on how they are observed by physicists. This is elementary quantum physics as expressed by an elementary linguistic device, the chiasmus; but in a poem from the "Morphemics" section of his volume, *Language*, Spicer adds a complex feature to the equation:

> Lew, you and I know how love and death matter
> Matter as wave and particle — twins
> At the same business (1975, p. 234).

Frederick Amrine (ed.) Literature and Science as Modes of Expression, 177–188.
© 1989 *Kluwer Academic Publishers, all rights reserved.*

inary apples in the metaphor 'apple cheeks', for instance. To the poet who takes the Newtonian universe for granted, real apples Scotch-taped to real cheeks would be an absurd literalization. To the Newtonian scientist, imaginary apples are pretty in poetry, but only the cheeks are 'real'. These distinctions, which assume that linguistic systems must slavishly obey the laws of the macrocosm, extend to the taboo that we must never mix our metaphors.

The world of the quantum, however, presents the observer with very different laws. Subatomic particles do not behave like objects in Newtonian space. Physicists perceive that the difficulties of interpretation which quanta present them with are not merely mathematical, they are linguistic as well. As Werner Heisenberg has observed, "All the words or concepts we use to describe ordinary physical objects, such as position, velocity, color, size, and so on, become indefinite and problematic if we try to use them on elementary particles" (1974, p. 114). The difficulty, however, cannot simply be pinned down to words alone. Some physicists wonder whether conventional distinctions between 'real' and 'unreal' — the heart of the matter of metaphor — always apply to the microcosm, where the question, "Are quanta real?" is taken seriously (Jauch).

Jack Spicer takes the same questioning stance toward the metaphors of poetry. The following poem from *Language* provides a good example of the poet's guerrilla strategy against metaphor:

Sporting Life

The trouble with comparing a poet with a radio is that radios don't develop scar-tissue. The tubes burn out, or with a transistor, which most souls are, the battery or diagram burns out replaceable or not replaceable, but not like that punchdrunk fighter in the bar. The poet

Takes too many messages. The right to the ear that floored him in New Jersey. The right to say that he stood six rounds with a champion.

Then they sell beer or go on sporting commissions, or, if the scar tissue is too heavy, demonstrate in a bar where the invisible champions might not have hit him. Too many of them.

The poet is a radio. The poet is a liar. The poet is a counter-punching radio.

mirror the laws of nature (John throws the ball to Robert). But language may disobey Newtonian laws — in syntax, grammar, metaphor — even as quanta do. Spicer observes that "We make up a different language for poetry/And for the heart — ungrammatical" (1975, p. 233). Even the "language of the heart" — everyday words and sentences which people speak to each other — fails when we "cannot quite make the sounds of love/The language/Has so misshaped them" (1975, p. 237).

Quantum poetics and quantum physics share another epistemological feature: both systems are dedicated to tearing down boundaries which the human mind has artificially imposed upon matter and language. Modern physicists even question the boundary between macrocosm and microcosm:

> ... we maintain that all our macroscopic bodies of classical physics are composed of atoms and elementary particles held together by forces of various kinds. There must therefore exist a boundary where the classical description ceases to have validity and the quantum properties become dominant. Now nobody knows the exact position of the boundary. Most people would agree that the experimental apparatus with which we execute the experiments and the computers with which we evaluate the data are on the classical side, and therefore behave according to the laws of classical physics. But between this input and output there is a system, like the photons ... which behaves quite differently from any classical system that we know. Thus, by setting a boundary somewhere, on one side of which things are classical and on the other side quantal, we cause almost insoluable problems of fundamental importance (Jauch, pp. 32—3).

Thus, for the physicist J. M. Jauch, the most hallowed boundary of all, the 'line' between the Newtonian world of classical mechanics, and the quantum realm, may not exist, at least in a configuration which makes sense to us at present.

Jack Spicer is equally suspicious of the artificial boundaries which are created by human discourse:

> Let us tie the strings on this bit of reality.
> Graphemes. Once wax now plastic, showing the ends. Like a red light.
> One feels or sees limits.
> They are warning graphemes but also meaning graphemes because without the marked ends of the shoelace or the traffic signal one would not know how to tie a shoe or cross a street — which is like making a sentence (1975, p. 240).

Graphemes, which along with morphemes and phonemes constitute the

> Sandy growls like a wolf. The space between him and his image
>   is greater than the space between me and my image.
> Throw him a honey-cake. Hell has been proved to be a series of
>   image.
> Death is a dog and Little Orphan Annie
> My own Eurydice. Going into hell so many times tears it
> Which explains poetry (1975, p. 226).

Here, Spicer uses two familiar comic strip characters to suggest that the "series of image" which "Hell has been proved to be" is a lie. Death and Hell are unmetaphorical; they are real to the speaker who has gone "into hell so many times". Thus, the poem punches through the pasteboard masks of its own metaphors: Sandy who growls like a wolf becomes Cerebus; Little Orphan Annie becomes Eurydice; both comic strip characters are always a step (a vehicle) away from the real. The "space" between them and their images is "greater" than the poet's because human beings do not suffer metaphorically. Human suffering "explains poetry" and also explains the speaker's suspicions of the very metaphors which appear in his poetry.

Spicer's deconstructions of metaphor are occasionally playful, as in the erotic baseball poems from *Book of Magazine Verse*:

> I would like to beat my hands around your heart.
> You are a young pitcher but you throw fast curve-balls, slow
>   fast-balls, change-ups that at the last moment don't change.
>   Junk
> The pitchers who are my age call it. And regret every forty
>   years of their life when they have to use them (1975, p. 256).

Spicer undercuts conventional metaphors of religion as well:

> God is a big white baseball that has nothing to do but go in a
>   curve or a straight line. I studied geometry in high school and
>   know that this is true.
> Given these facts the pitcher, the batter, and the catcher all look
>   pretty silly . . .

The speaker concludes,

>   . . . Off Seasons
> I often thought of praying to him but could not stand the
>   thought of that big, white, round omnipotent bastard.

C. G. Jung has written, "cannot be localized in space or ... space is relative to the psyche". Jung elaborates:

Synchronistic phenomena prove the simultaneous occurrence of meaningful equivalences in heterogeneous, causally unrelated processes; in other words, they prove that a content perceived by an observer can, at the same time, be represented by an outside event, without any causal connection (p. 518).

In a whimsical poem from the "Intermissions" section of *Language*, Spicer suggests that the poet also participates in Jungian acausality:

Where is the poet? A-keeping the sheep
A-keeping the celestial movement of the spheres in a long,
    boring procession
A-center of gravity
A-(while the earthquakes of happiness go on inside and outside
    his body and the stars in their courses stop to notice)
Sleep (1975, p. 230).

Like William Blake, perhaps, who hated Newtonian cosmology, and felt that "The stars were in the heavens because man's imagination saw them there" (Kazin, p. 4), Jack Spicer takes seriously the possibility that the physical universe, and man's perception of it, are not two separate systems.

For David Bohm in a quantum context, and for Jung in a Newtonian one, human beings live in a world of processes, not of watertight distinctions between big and small, self and world, mind and matter.[2] In epistemological harmony with both Bohm and Jung, Jack Spicer seeks to restructure the devices of poetry in order to pay proper tribute in language to such a world. Perhaps the most interesting feature of Spicer's quantum poetics is the deconstructions of metaphor which appear in his later work.

2

Of all poetical conventions, the metaphor is particularly 'user friendly' in a Newtonian universe, where distinctions between here/there and either/or are taken for granted. Indeed, the two working parts of a classic metaphor, tenor and vehicle, reinforce epistemological distinctions between 'real' and 'fanciful', between worldly cheeks and imag-

Spicer's view of what he simply calls "poetry", which mysteriously "comes along after the city is collected", resembles Michel Foucault's description of the Other, or the ideal "unthought" which always escapes human attempts to grasp it:

> Man and the unthought are, at the archaeological level, contemporaries. Man has not been able to describe himself as a configuration in the episteme without thought at the same time discovering, both in itself and outside itself, at its borders yet also in its very warp and woof, an element of darkness, an apparently inert density in which it is embedded, an unthought which it contains entirely, yet in which it is also caught. The unthought (whatever name we give it) is not lodged in man like a shrivelled-up nature or a stratified history; it is, in relation to man, the Other: the Other that is not only a brother but a twin, born, not of man, nor in man, but beside him and at the same time, in an identical newness, in an unavoidable duality (p. 326).

When Spicer describes as an "unavoidable metaphor" the relationship between the poet and the "ghost of the poem", or the perfect voice which he seeks (in vain) to make his own, he recalls Foucault's "unavoidable duality". What both men are suggesting is that, in the mysterious symbiosis between human and Other, the attraction is mutual. For Spicer, poetry becomes

> ... A silver wire which reaches from the end of the beautiful as if elsewhere. A metaphor. Metaphors are not for humans.
>
> The wires dance in the wind of the noise our poems make. The noise without an audience. Because the poems were written for ghosts.
>
> The ghosts the poems were written for are the ghosts of the poems. We have it second-hand. They cannot hear the noise they have been making.
>
> Yet it is not a simple process like a mirror or a radio. They try to give us circuits to see them, to hear them. Teaching an audience (1975; p. 170).

The "ghosts", or the Other which the poet intuits, struggle silently ("They cannot hear the noise they have been making") in the webs of closure ("the imposition of an order" which metaphor is) and try "to give us circuits to see them, to hear them". The "circuits" represent the

And those messages (God would not damn them) do not even
know they are champions (1975, p. 218).

Like a Cubist painting, "Sporting Life" offers the spectator/reader a familiar image, the neoclassical metaphor of the poet as transmitter, brought up to date as a radio. Then, as in Cubism, the poem proceeds to tear its subject matter to pieces — in the case of "Sporting Life", the subject matter includes metaphors for the poet and the speaker's avowed skepticism of those metaphors. In "Sporting Life", Spicer's deconstructions of metaphor occur on three levels. First, the poet plants a seed of doubt in the reader's mind by pointing out a flaw in the analogy between poet and radio which extends throughout the poem: "radios don't develop scar tissue". Second, the poet further undercuts the metaphor by breaking the rules and mixing poet as radio with poet as fighter: "The poet is a counterpunching radio". Finally, the speaker shatters both metaphors by telling the reader, "The poet is a liar". What the poet "is", in other words, can only be expressed metaphorically, and thus falsely. Only the "messages" which come from an outside source unknown to the poet ("God would not damn them") are real. Spicer's version of Epimenides' Cretan paradox, "The poet is a liar", forces the reader into a thicket of impossible questions, among them: Is the poet (speaker) a liar who is telling the truth through metaphor? Or is he a trustworthy speaker who is lying through metaphor? These questions don't make sense; neither do conventional metaphors, which, according to the metaphorical poet, always allow some truth ("scar tissue") to escape. For Spicer, these deconstructions of metaphor are not clever apparatuses to mirror the content of the poem; the deconstructions are the content of the poem.

Spicer's attacks on metaphor can be obscene:

For you I would build a whole new universe around myself.
    This isn't shit it is poetry. Shit
Enters into it only as an image (1975, p. 225).

Or:

In my heart, as Verlaine said, I can hear the little sound of it
    raining
Not an indian sign. But real unfucking rain (1975, p. 239).

More often, the attacks are poignant:

"infinitely small vocabulary" of nature continue twenty years after Spicer's death. Meanwhile, the speakers of Spicer's poems still argue with their own metaphors, argue with the ghosts of the poems, who

>   ... won't come through. Nothing comes through. The death
>
>   Of every poem in every line
>
>   The argument con-
>   tinues (1975, p. 171)

NOTES

[1] The physicist David Bohm proposes what he calls a "rheomode", or a system of radical grammar and syntax which attempts to bring language more into line with physical reality (ch. 2).
[2] See also Owen Barfield's seminal study, *Saving the Appearances*.

BIBLIOGRAPHY

Barfield, O.: *Saving the Appearances*, Faber and Faber, London, 1957.
Blaser, R.: 'The Practice of Outside', in J. Spicer, *The Collected Books of Jack Spicer* (ed. by R. Blaser), Black Sparrow Press, Los Angeles, 1975.
Bohm, D.: *Wholeness and the Implicate Order*, Routledge and Kegan Paul, London, 1980.
Davidson, M.: 'Incarnation of Jack Spicer. Heads of the Town Up to the Aether', *Boundary 2* **6** (Fall, 1977) pp. 103—44.
Foucault, M.: *The Order of Things*, Pantheon, New York, 1970 (qtd. in R. Blaser).
Heisenberg, W.: *Across the Frontiers*, Harper and Row, New York, 1974 (qtd. in G. Zukav, *The Dancing Wu Li Masters*, Bantam Books, New York, 1979, p. 21).
Jauch, J.: *Are Quanta Real?*, Indiana Univ. Press, Bloomington, 1973.
Jung, C.: *The Portable Jung* (ed. by J. Campbell), Penguin, New York, 1971.
Kazin, A.: 'Introduction' in W. Blake, *The Portable Blake* (ed. by A. Kazin), Viking, New York, 1968.
Spanos, W.: 'Jack Spicer's Poetry of Absence: An Introduction', *Boundary 2* **6** (Fall, 1977) pp. 1—2.
Spicer, J.: 'From the Vancouver Lectures', *Caterpillar 12* (July, 1970) p. 206.
Spicer, J.: *The Collected Books of Jack Spicer* (ed. by R. Blaser), Black Sparrow Press, Los Angeles, 1975.

California State University, Bakersfield

Yet he's there. As the game follows rules he makes them.
I know
I was not the only one who felt these things (1975, p. 258).

Instead of God as Light, or God as a bearded patriarch, Spicer compares God to a baseball. What the speaker "could not stand", of course, was the "thought of" God as metaphor, even though human beings are doomed always to think of Him metaphorically. Once the baseball disappears in the last two lines (along with the geometric logic of Newtonian laws of motion), the poem becomes a simple statement of faith.

3

Spicer draws a parallel between the matter of language and the language of matter in a series of lectures he gave at the Vancouver Poetry Festival in the summer of 1965. He finds degrees of complexity in language analogous to "the most simple particles and the more complicated particles in chemistry-physics" (1970, p. 210). But as Spicer might have added, the trouble with comparing words with particles, as if the poem were merely a sort of linguistic particle accelerator, is that neither words nor particles develop scar tissue. Put another way, for Spicer, deconstruction of language, and of metaphor in particular, becomes deconstruction of self. Every metaphor is a mask: so are the poets who make them:

> What I am, I want, asks everything of everyone, is by degrees a ghost. Steps down to the first metaphor they invented in the underworld (pure and clear like a river) the insight. As a place to step further (1975, p. 182).

"A place to step further" means that every tenor in poetry is a mask (a disguised vehicle) for another tenor. That tenor is a mask (a disguised vehicle) for another tenor, and so forth. Similarly, every human being is a mask for a ghost. For Spicer, 'ghost' may also be a metaphor for the Other, the hidden voice or shadow which every poet desires to capture in the perfect poem. If the speaker of the poem is "by degrees a ghost", then no boundary exists between him and the Other. The Speaker may be "real" (tenor) and the Other may be "imaginary" (vehicle). Or the reverse may be true:

> Poetry comes along after the city |of poets| is collected. It recognizes *them* as a metaphor. An unavoidable metaphor. Almost the opposite (1975, p. 175).

INDEX OF NAMES

Cadden J. xiii, xxiii
Cajal, R. y 39, 48
Campbell, J. 188
Canning, G. 55
Carlisle, E. xvi, xxiii
Carlson, E. 141–2, 149–50
Carter, S. xvii, xx
Cassirer, E. 97
Castillejo, D. 111
Cather, W. 171, 175
Cesbron 56
Cheyne, G. 39, 41, 43, 55
Cicero 96
Coetzee, J. 24–5
Cohen, I. 5, 25, 58
Cohen, R. ix, xi
Coleridge, S. 43, 57, 78, 129–30, 139
Compere, D. 25
Condillac, E. xvii, 68–70, 77–99
Condorcet, M. de 45, 58
Conti 102
Coventry, C. 143, 150
Crane, H. xvii, 6, 25
Croll, M. 31
Crosland, M. 107, 110–11
Crowe, M. 6, 25
Cullen, W. 41, 44, 46, 57

D'Alembert, J. 77, 79
Daniel 105
Dante xiv
Darwin, C. xvi, xvii, 10–12, 18, 25, 30
Davidson, M. 177, 188
Debus, A. 57–8, 99, 105–6, 110–11
Decartes, R. 29, 39, 56, 65, 77, 133
Dee, J. 104–5, 110
Defoe, D. 10, 24–5, 41
Deleuze, G. 4, 25, 91, 97
de Lillo, D. 165
de Man, P. 83, 97, 130, 134, 139
Derrida, J. 8, 25, 83, 96–97, 122, 133, 139, 187
Deutsch, V. 97
Dewhurst, K. 36, 57–8
D'Holbach, P. 55
Dick, H. 111
Dickinson, L. 1

Diderot, D. 41, 55, 77, 79
Dobbs, B. 99, 110–11
Donato, E. 85, 88, 97
Donne, J. 6, 25
Dos Passos, J. xvii, xxii, 163–4, 169, 172–5
Dover, T. 54
Dowdell, V. 98
Drake, S. 111
Dreiser, T. 170
Drucker, P. 166

Eade, J. 59
Eastman, G. 161
Edison, T. 156, 161, 163
Eggers, T. 54
Eibl, K. 72, 79
Eichner, H. 137, 139
Einstein, A. 177–8
Eisenhower, D. 164
Elkana, Y. ix, xi
Eliot, T. xvii, 5–6, 25, 165–6, 171
Emerson, R. 156
Empson, W. 40
Engell, J. 139
Epimenides 183
Euclid 104
Evans, W. 174–5

Faulkner, W. 162
Feingold, M. xi
Fekete, J. 96–7
Fernandes, S. 96–7
Fernel, J. 56
Ferrand, E. 54
Feyerabend, P. 59, 132, 138, 140
Fichte, J. 118, 120, 123–4, 127, 131–5
Fielding, H. 41
Figuier, L. 45, 58
Fish, S. 3, 25
Fitzgerald, F. 162
Flemyng, M. 43, 55
Forbes, E. 13, 25
Ford, H. 156, 161
Fornet-Betancourt, R. 96
Forsyth, P. 26
Fosgate, B. 144, 150
Foucault, M. 29, 36–7, 58, 82, 96–7, 122,

poet's desire to write the perfect poem. Without that desire, the "ghosts" disappear from the poem. The poet is left with mere words, which, as Spicer says elsewhere, "are what sticks to the real . . . They are what we hold on with, nothing else. They are as valuable in themselves as rope with nothing to be tied to" (1975, p. 309).

The deconstruction of the poet whose own metaphors turn against him, thus raising the terrible possibility that it is *he* who "was never real" (1975, p. 80), and the "ghosts" who recognize *him* as a metaphor, takes its toll. At times Spicer seems near despair:

> This is the crab-god shiny and bright
> who sunned by day and wrote by night
> And lived in the house that Jack built.
> This is the end of it, very dear friend, this
> is the end of us (1975, p. 233).

In the poem (the "house"), the poet's self ("Jack") is as metaphorical as "the crab-god" which "Jack" selects as a metaphor for himself. Conversely, "Jack" may be a metaphor for "the crab-god". For the poem reveals that a construction of metaphor is merely a projection of a similarly constructed human self; the infinitely evasive tenor is a projection of the infinitely evasive being of the speaker. The deconstructed poet ("Jack") and his deconstructed metaphor ("the crab-god") therefore end simultaneously as the poem ends: "this/is the end of us". As he lay dying of acute alcoholism at forty, Jack Spicer's last words were, "My vocabulary did this to me" (1975, p. 325).

And yet the poet is not entirely defeated. The poems remain as testaments to the struggle of every poet to find the perfect poem: a place in language to step no further. Before Jacques Derrida's milestone paper on the Discourses of Human Sciences at Johns Hopkins in 1966, Jack Spicer had anticipated post-structuralism by saying that "the perfect poem has an infinitely small vocabulary" (1975, p. 25). What he means, of course, is that the perfect poem cannot be written; it can only be uttered by the still, small intuited voice of the Other, and never in human language. Nevertheless, it is the shadowy presence of the Other ("the crab-god") which makes language infinitely beautiful.

Jack Spicer's work, then, is in epistemological harmony with both the post-structuralists and with quantum physicists, whose searches for the

INDEX OF NAMES

John of Garland 104, 110
Johnson, B. 97
Johnson, J. xxiv
Jolly, A. 18, 26
Jones, R. 31
Jordanova, L. xvi, xviii, xxiii, 2, 26, 96–7
Josten, C. 110–11
Jowett, B. 27
Jung, C. 8–9, 26, 48, 181, 188

Kamensky, G. xxiv
Kant, I. xvii, xix, 33, 43, 56, 65, 72, 82–96, 98, 116–8, 120, 122–5, 127, 131, 133, 137
Kapitza, P. 129, 140
Kapp, E. 153
Kargon, R. 77, 79
Kaser, D. 151
Kasson, J. 158, 167
Kazin, A. 181, 188
Keller, E. xx, xxiii
Kelly, E. 103, 111
Kern, S. 21
Kevles, D. 166–7
Kimbal, J. 169–70, 175
King, H. 56
Kipperman, M. xvi–xxiii
Knapp, J. 150
Knoespel, K. xvii–xviii, 110, 112
Koch, R. xvii, xix
Koerner, J. 164, 167
Koestler, A. 42
Kohn, D. 25
Kouwenhoven, J. 155, 167, 175
Kuhn, T. 91, 97–8, 103, 111, 166–7, 132, 140
Kuzniar, A. 130, 134–5, 140

Lacquer, T. 31, 58
Lafayette, M. de 44
Lamarck, J. 30
Laurens, A. du 32, 34, 54
Lavoisier, A.-L. 64–5, 72, 79, 107–8
Lawson, T. xxiii
Layton, E. 157, 166–7
le Doeuff, M. 24–5
Lee, J. xvi, xxiii
Leff, M. 96, 98

Lem, S. 24, 26
Leibniz, G. 45, 91, 96, 115, 133
Lesch, J. 59
Lesson, M. 12
Levin, C. 96
Levine, G. x, xi, xvi, xx, xxii–xxiii, 2, 25–6
Lévi-Strauss, C. 22–3, 26
Lewis, S. 159–60
Lichtenberg, G. 62–3, 116
Linge, D. 97
Link, H. 131, 140
Lippmann, W. 161
l'Isle-Adam, V. de 66
Livingston, P. x, xi
Locke, J. 30, 36
Lorry, A. 54
Lucas, P. 98
Lukács, G. 126
Lyotard, F. 110–11

Mabbott, T. 146, 151
MacArthur, R. 17, 26
Mach, E. xvii, xx, 61–4, 67–8, 76, 79
Macquer, P. 107–8
Mahoney, D. 129, 139–40
Maier(us), M. 2, 100, 103, 110
Mallock, W. 14, 26
Mandelbrot, B. 9, 24, 26
Mandeville, B. xv, 38, 43, 54
Manguel, A. 26
Mannheim, R. 97
Marat, J. 55
Marivaux, P. 70–71, 78–9
Markus, G. xxi, xxiii
Marquis, A. 166–7
Martin, G. 96, 98
Marvin, U. 26
Marx, K. 37, 83
Marx, L. 156–7, 167
Matson, R. 56, 59
Maxwell, C. xiv
Maxwell, G. 59
McGrath, W. 56, 59
McLeod, I. 97
Medawar, P. 1, 26
Mehlman, J. 24, 26
Mendeléef, D. xiv, xxiii

INDEX OF NAMES

Abriscosoff, G. 35, 56, 58
Adams, H. 157, 164–5
Adair, J. 55
Adorno, T. 83, 132, 139
Agee, J. xvii, 160, 169, 172–5
Alberti, M. 55
Albertsen, L. 126–7
Allen, W. 158
Allnutt 150
Allott, K. 24
Amrine, F. xxiii
Anderson, W. xviii, xxiii, 107–8, 111
Apel, K.-O. 126–7
Arnold, M. xiii, xvii, 12, 24
Ashmole, E. 101, 103–4, 110–11
Austen, J. 41, 43–4, 55

Bachelard, G. 15, 25, 45, 91, 97
Bacon, F. 4, 10, 33, 65, 77, 104, 111
Baillager 150
Bakhtin, M. 41, 135, 139
Ball, J. 149, 151
Ballantyne, R. 25
Barfield, O. 188
Bar-Hillel, Y. xxiii
Barricelli, J.-P. xxiii
Barth, J. 164
Barton, W. 97
Bate, W. 139
Battaglia, S. 100, 111
Baxter, R. 54
Baynes, D. [Kinneir] 55
Bazerman, C. xvii, xviii, 110
Beauchene, E. de 55
Becker, H. 96
Beer, G. ix, xvi–xix, xxiii, 2–3, 6, 24–5
Bell, C. 39
Bennington, G. 97

Berkenhout, J. 55
Bernard, C. 66
Bhabba, H. 25
Bianchini 56
Bjurman, G. 150
Black, M. xvi, xxiii
Blackmore, R. 54
Blair, C. 89, 97
Blake, W. 40, 47, 181
Blaser, R. 178, 180, 188
Böhme, J. 40
Boerhaave, H. 45
Bohm, D. 94, 97, 180–81, 188
Bonnet, C. 45
Bottomley, F. 31, 58
Bourke-White, M. 174
Boyle, R. 43–5, 58, 105, 111
Brain 56
Brecht, B. 72, 74–5, 79
Breton, N. 54
Breuer, J. 38
Brigham, A. 144, 151
Bright, T. 32–4, 54
Brissenden, R. 59
Brooks, V. 175
Brostowin, P. xiii, xxiii
Brown, J. 44, 46, 57, 131–2, 138
Browning, R. 14, 25
Bruno, G. 46
Bruns, G. 96–7
Buchdahl, G. 85, 97
Buck, P. 163
Buffon, C. de [G. Leclerc] 69–72, 78–9
Bullen, B. 25
Burrow, J. 25
Burton, R. 32–4, 37, 39
Bynum, P. xvii, xviii
Byron, G. 13, 25

189

INDEX OF NAMES

Rudwick, M. 3–4, 27
Rush, B. 141–3, 146, 149–51
Russell, B. 59

Sade, M. de 46
Saint-Pierre, J. 10, 27
Saliba, D. 141, 148, 151
Saltonstall, W. 13–14, 27
Santayana, G. 162
Sarton, G. 53, 58–9
Saussure, F. de 84
Savioz, R. 45, 60
Schachterle, L. xiv, xxiii–xxiv
Schatzberg, W. xvii, xxiv
Schelling, F. 116, 135, 137
Schiller, F. 116, 125
Schlanger, J. 77, 79
Schlegel, A. 118
Schlegel, F. 118, 120, 130
Schmidt, A. 66
Schöne, A. 76–7, 80
Scholes, R. 52, 60
Schrader, G. 96, 98
Schumpeter, J. 161, 167
Schuster, J. xvi, xxiv
Schwartz, W. 98
Scott, H. 158
Searle, J. 64, 80
Seward, W. 144
Sextus Empiricus 88
Shakespeare, W. 6, 20, 27
Shaffer, E. xxii, xxiv
Shannon, C. 164, 167
Shapin, S. xvi, xxiv
Shapiro, K. 175
Shelley, M. 66
Shelly, P. 130
Sherlock, M. 55
Shulman, R. 146, 151
Shumaker, W. 112
Sigmond, G. 147, 150–51
Silverman, H. 96
Singer, C. 154, 167
Singer, I. 156
Skultans, V. 151
Slade, J. xvi–xvii, xxii, xxiv, 158, 167
Slaughter, M. 110, 112

Slusser, G. xvi, xxiii
Smith, A. 55
Smith, D. 55
Smith, H. 55
Smith, N. 98
Smith, W. 55
Smollett, T. 41
Smyth, R. 96, 98
Sontag, S. 166, 168
Soranus of Ephesus 35
Spanos, W. 177, 188
Spicer, J. xvii, xx, 177–88
Spiller, R. 166, 168
Spinoza, B. 81, 89, 134
Spivak, G. 25
Stahl, G. 40
Steele, R. 44
Stein, G. 21
Steinberg, L. xxi, xxiv
Steiner, R. 126
Steinman, L. 160, 168
Steinmetz, C. 163
Stephens, J. 112
Sterne, L. 34, 41–4, 55
Stevens, W. 160
St. John, J. 20, 26
Struik, D. 154, 168
Stukeley, W. 54
Sydenham, T. 36–8, 54
Synge, E. 54
Swedenborg, E. 40
Swift, J. 4, 10, 27, 34, 161

Taylor, F. 158, 163
Taylor, J. 33, 60
Temkin, O. 57
Tennyson, A. 14, 27
Thackeray, W. 6, 27
Thompson, D. 96
Thomson, H. 55
Thomson, J, 57, 60
Tichi, C. 158–60, 166, 168, 175
Tissot, A. 55, 57
Todorov, T. 102, 109–10, 112
Tollius, J. 105
Tomlinson, H. 97
Toulmin, S. 23, 27

INDEX OF NAMES

186, 188
Frame, J. 15, 26
Francis of Assisi 35
Freeman, W. 144, 149–51
French, R. 67–8
Freud, S. 8–9, 34, 38, 48, 117
Frye, N. 44, 58

Gadamer, H.-G. 96–7
Gaddis, W. 165
Gaier, U. 139–40
Galen 56
Galileo 50, 106, 111
Gallagher, C. 31, 58
Gantt, H. 158
Gardam, J. 16, 24, 26
Garvin, H. xxiv
Gauthier, J. 96
Gay, P. 43
Gehlbach, F. 17, 26
Ghisalberti, F. 110–11
Gibaldi, J. xxiii
Gilman, S. xi
Glanville, J. 1
Godwin, W. 42
Goethe, J. 57, 66, 76, 116, 125, 129–30
Goldberg, R. 170
Golding, W. 10, 26
Gomez–Muller, A. 96
Graña, C. 155, 166
Greene, J. 154, 166
Gregory, R. 55, 58
Guadalupi, G. 26
Guattari, F. 4, 25
Guffey, G. xvi, xxiii
Guthke, K. 57–8

Habberjam, B. 97
Habermas, J. 74–5, 79, 96, 121, 126–7
Hacking, I. xx, xxiii
Hadfield 151
Hall, B. 151
Haller, A. 41, 44–8, 55, 57–9
Hannah, R. 131, 140
Hannaway, O. xviii, xxiii
Hardy, T. 17
Harris, L. 17, 26

Harris, N. 161, 167
Harrison, T. 19, 26
Hart, H. 175
Harte, W. 54
Hartman, R. 98
Hartung, H. 66, 79
Haslam, J. 143, 151
Hawes, D. 27
Hayles, K. xvi, xxiii
Hayles, N. xi
Hayward, J. 25
Heath, J. xiv
Heidegger, M. 83, 96–7
Heine, H. 126
Heisenberg, W. 174, 178, 182, 188
Heissenbüttel, H. 66
Helvetius, M. 55
Hemingway, E. 6, 162–163, 171
Hesse, M. x, xvi, xxiii
Hilton, N. 47, 59
Hindle, B. 154, 167
Hippocrates 35, 56
Hochberg, H. 58–9
Hodges, D. 56, 59
Hölderlin, F. 130
Holton, G. 3, 18, 26, 91, 97
Homer 10
Hooker, J. 10, 12, 26
Hoover, H. 158
Horkheimer, M. 132, 139
Howarth, W. 151
Howe 56
Hughes, T. 161, 167
Hunter, J. 44
Husserl, E. 83, 97
Huxley, A. xiii–xv, xxiii, 4, 26, 30
Huxley, T. xiii–xiv

Ihde, D. 96
Isenflamm, J. 55

Jack, I. 25
Jackson, S. 56, 59
Jameson, F. 110–11
Jauch, J. 179, 182, 188
Jefferson, T. 154, 164
Jennis 104

INDEX OF NAMES

Meredith, J. 98
Milton, J. 105
Molnár, G. xvii, xx, 119, 127, 130–31, 133–6, 140
Molyneux, W. 71
Montesquieu, C.-L. 45
Mooney, T. 165
Moore, M. 160, 170
More, T. 10, 19–20, 26
Morgan, J. 161–2
Morton, P. 166–7
Moser, W. xvii, xx
Multhauf, R. 155, 167
Mulkay, M. xvi, xxiv
Murhard, F. 131
Musil, R. 66, 76–7
Myers, V. 41, 59

Nachelmans, G. 85, 98
Nagel, T. 31, 39, 59
Nancy, J.-L. 96, 98
Napoleon 118
Neubauer, J. xiv, xvi–xvii, xix, xxiv, 57, 59, 140
Newton, I. 45–46, 85, 93–4, 105, 110–11, 177–82, 185
Nicholson, M. xiv, xxiv, 30
Nietzsche, F. 89, 118, 122, 132, 139
Norris, F. 170
Norton, T. 103
Novalis [F. von Hardenberg] xvii, xix–xx, 77–9, 116, 118–21, 123–6, 129–40

O'Brien, E. 163, 167
Olson, C. 9
Orvell, M. xvii, xxii
Ovid 13

Padre Pio 35
Paracelsus 40
Paradis, J. xi
Pearson, B. 26
Peat, D. 97
Peirce, C. 85, 126
Pernety, A.-J. 105
Peter, K. 126–7
Peterfreund, S. xvi–xvii, xxiv

Pharies, D. 98
Pindar 159
Pinel, P. 142, 149
Platerus, F. 32, 34, 54
Plato 7, 27, 170
Plotinus 139
Plum, F. 39, 45, 59
Poe, E. xvi–xviii, 141, 143–51
Pois, C. le 38, 54
Pomme, P. 55
Pool, I. 167
Popper, K. 138, 140
Porat, M. 157, 164, 167
Porter, R 30–31, 59
Pound, E. 171, 175
Pratt, M. 58–9
Prichard, J. 142–3, 145, 149, 151
Priestley, J. 42
Pynchon, T. 164–5

Quinn, A. xi
Quintilian 88, 96

Raban, J. 20, 26
Rabelais, F. 10, 41
Rather, L. 56, 59
Ray, I. 144
Reeves, N. 57–8
Reid, J. 147, 151
Reilly, J. 150–151
Reingold, N. 154, 167
Reiss, T. 58–9
Richards, I. 3, 5, 26
Richelieu 30
Ricks, C. 27
Ripa 102
Ripley, G. 100–101, 110
Roberts, L. xi
Robinson, A. 146, 151
Robinson, N. 54
Rolfe, B. 144, 150
Rorty, R. 57
Rousseau, G. x, xi, xiii, xv–xix, xxii, xxiv, 29, 31, 41, 57, 59
Rousset, J. 78–9
Rowley, W. 55
Rudolph, G. 57, 59

20. Kenneth F. Schaffner and Robert S. Cohen (eds.), *Proceedings of the 1972 Biennial Meeting, Philosophy of Science Association.* 1974
21. R. S. Cohen and J. J. Stachel (eds.), *Selected Papers of Léon Rosenfeld.* 1978.
22. Milic Čapek (ed.), *The Concepts of Space and Time. Their Structure and Their Development.* 1976.
23. Marjorie Grene, *The Understanding of Nature, Essays in the Philosophy of Biology.* 1974.
24. Don Ihde, *Technics and Praxis. A Philosophy of Technology.* 1978.
25. Jaakko Hintikka and Unto Remes, *The Method of Analysis. Its Geometrical Origin and Its General Significance.* 1974.
26. John Emery Murdoch and Edith Dudley Sylla, *The Cultural Context of Medieval Learning.* 1975.
27. Marjorie Grene and Everett Mendelsohn (eds.), *Topics in the Philosophy of Biology.* 1976.
28. Joseph Agassi, *Science in Flux.* 1975.
29. Jerzy J. Wiatr (ed.), *Polish Essays in the Methodology of the Social Sciences.* 1979.
30. Peter Janich, *Protophysics of Time.* 1985.
31. Robert S. Cohen and Marx W. Wartofsky (eds.), *Language, Logic and Method.* 1983.
32. R. S. Cohen, C. A. Hooker, A. C. Michalos, and J. W. van Evra (eds.), *PSA 1974: Proceedings of the 1974 Biennial Meeting of the Philosophy of Science Association.* 1976.
33. Gerald Holton and William Blanpied (eds.), *Science and Its Public: The Changing Relationship.* 1976.
34. Mirko D. Grmek (ed.), *On Scientific Discovery.* 1980.
35. Stefan Amsterdamski, *Between Experience and Metaphysics. Philosophical Problems of the Evolution of Science.* 1975.
36. Mihailo Marković and Gajo Petrović (eds.), *Praxis, Yugoslav Essays in the Philosophy and Methodology of the Social Sciences.* 1979.
37. Hermann von Helmholtz, *Epistemological Writings. The Paul Hertz/Moritz Schlick Centenary Edition of 1921 with Notes and Commentary by the Editors.* (Newly translated by Malcolm F. Lowe. Edited, with an Introduction and Bibliography, by Robert S. Cohen and Yehuda Elkana). 1977.
38. R. M. Martin, *Pragmatics, Truth, and Language.* 1979.
39. R. S. Cohen, P. K. Feyerabend, and M. W. Wartofsky (eds.), *Essays in Memory of Imre Lakatos.* 1976.
40. B. M. Kedrov and V. Sadovsky. *Current Soviet Studies in the Philosophy of Science.* Forthcoming.
41. M. Raphael, *Theorie des Geistigen Schaffens auf Marxistischer Grundlage.* Forthcoming.
42. Humberto R. Maturana and Francisco J. Varela, *Autopoiesis and Cognition. The Realization of the Living.* 1980.
43. A. Kasher (ed.), *Language in Focus: Foundations, Methods and Systems. Essays Dedicated to Yehoshua Bar-Hillel.* 1976.
44. Trân Duc Thao, *Investigations into the Origin of Language and Consciousness.* (Translated by Daniel J. Herman and Robert L. Armstrong; edited by Carolyn

INDEX OF NAMES

Tournier, M. 24, 27
Trachtenberg, A. 161, 168
Tracy, D. de 40, 60
Trillat 56
Turner, B. 31, 60
Twain, M. 170

Valentinus, B. 100, 110
Van Buren, J. 144
Van Lennep, J. 110, 112
Van Nest family 144, 148, 150
Van Nest, J. 151
Veblen, T. 158, 160, 168
Veith, I. 35, 56, 60
Verlaine, P. 183
Verne, J. 27
Vertov, J. 27
Vesalius, A. 56
Vickers, B. 56, 60, 112

Waite, A. 112
Waite, R. xxiv
Wajeman 56
Wallace, A. 10, 12, 27
Ward, S. 106
Wartofsky, M. ix, xi
Watson, A. 147, 151
Weaver, W. 164, 167
Weber, B. 25
Webster, D. 150–51
Webster, J. 106
Wegener, N. xvii, 7, 9, 27

Weightman, D. 26
Weightman, J. 26
Weininger, S. J. ix, xi, xvi, xxiv
Weisskopf, V. 39, 45
Wells, H. 12
Westfall, R. 110, 112
Wharton, F. 143, 152
Whitman, W. 156, 169, 172–3, 175
Whytt, R. 41, 45, 48, 55, 57
Williams, W. 160, 171, 175
Williamson, M. 16–17, 27
Willis, T. 36, 38, 54
Wilson, E. 17, 26
Wilson, T. 7, 27
Winstanley, D. 56, 60
Wittig, M. 14, 27
Wood, J. 145, 151
Woodcock, J. xv, xxv
Woodward, S. 144, 146, 152
Woolf, V. 2, 20, 27
Woolgar, S. xvi, xxv
Wordsworth, W. 46, 57, 130, 140
Wubnig, J. 98
Wyss, J. D. 27
Wyss, J. R. 27

Yeo, R. xvi, xxiv
Young, R. 3, 27

Zola, E. 66, 72–4, 80
Zukav, G. 188

82. R. S. Cohen and M. W. Wartofsky, *Physical Sciences and History of Physics*. 1984.
83. E. Meyerson, *The Relativistic Deduction*. 1985.
84. R. S. Cohen and M. W. Wartofsky, *Methodology, Metaphysics and the History of Sciences*. 1984.
85. György Tamás, *The Logic of Categories*. 1985.
86. Sergio L. de C. Fernandes, *Foundations of Objective Knowledge*. 1985.
87. Robert S. Cohen and Thomas Schnelle (eds.), *Cognition and Fact*. 1985.
88. Gideon Freudenthal, *Atom and Individual in the Age of Newton*. 1985.
89. A. Donagan, A. N. Perovich, Jr., and M. V. Wedin (eds.), *Human Nature and Natural Knowledge*. 1985.
90. C. Mitcham and A. Huning (eds.), *Philosophy and Technology II*. 1986.
91. M. Grene and D. Nails (eds.), *Spinoza and the Sciences*. 1986.
92. S. P. Turner, *The Search for a Methodology of Social Science*. 1986.
93. I. C. Jarvie, *Thinking about Society: Theory and Practice*. 1986.
94. Edna Ullmann-Margalit (ed.), *The Kaleidoscope of Science*. 1986.
95. Edna Ullmann-Margalit (ed.), *The Prism of Science*. 1986.
96. G. Markus, *Language and Production*. 1986.
97. F. Amrine, F. J. Zucker, and H. Wheeler (eds.), *Goethe and the Sciences: A Reappraisal*. 1987.
98. Joseph C. Pitt and Marcella Pera (eds.), *Rational Changes in Science*. 1987.
99. O. Costa de Beauregard, *Time, the Physical Magnitude*. 1987.
100. Abner Shimony and Debra Nails (eds.), *Naturalistic Epistemology: A Symposium of Two Decades*. 1987.
101. Nathan Rotenstreich, *Time and Meaning in History*. 1987.
102. David B. Zilberman (ed.), *The Birth of Meaning in Hindu Thought*. 1987.
103. Thomas F. Glick (ed.), *The Comparative Reception of Relativity*. 1987.
104. Zellig Harris et al., *The Form of Information in Science*. 1987
105. Frederick Burwick, *Approaches to Organic Form: Permutations in Science and Culture*. 1987.
106. M. Almási, *Philosophy of Appearances*. Forthcoming.
107. S. Hook, W. L. O'Neill, and R. O'Toole, *Philosophy, History and Social Action. Essays in Honor of Lewis Feuer*. 1988.
108. I. Hronszky, M. Fehér, and B. Dajka (eds.), *Scientific Knowledge Socialized. Selected Proceedings of the Fifth Joint International Conference on History and Philosophy of Science Organized by the IUHPS, Veszprém, 1984*. Forthcoming.
109. P. Tillers and E. D. Green (eds.), *Probability and Inference in the Law of Evidence. The Uses and Limits of Bayesianism*. 1988.
110. E. Ullmann-Margalit (ed.), *Science in Reflection. The Israel Colloquium: Studies in History, Philosophy, and Sociology of Science*. 1988.
111. K. Gavroglu, Y. Goudaroulis, and P. Nicolacopoulos (eds.), *Imre Lakatos and Theories of Scientific Change*. 1989.
112. Barry Glassner and Jonathan D. Moreno (eds.), *The Qualitative-Quantitative Distinction in the Social Sciences*. 1989.
113. K. Arens, *Structures of Knowing: Psychologies of the Nineteenth Century*. 1989.

BOSTON STUDIES IN THE PHILOSOPHY OF SCIENCE

Editors:

ROBERT S. COHEN and MARX W. WARTOFSKY
(Boston University)

1. Marx W. Wartofsky (ed.), *Proceedings of the Boston Colloquium for the Philosophy of Science 1961–1962*. 1963.
2. Robert S. Cohen and Marx W. Wartofsky (eds.), *In Honor of Philipp Frank*. 1965.
3. Robert S. Cohen and Marx W. Wartofsky (eds.), *Proceedings of the Boston Colloquium for the Philosophy of Science 1964–1966. In Memory of Norwood Russell Hanson*. 1967.
4. Robert S. Cohen and Marx W. Wartofsky (eds.), *Proceedings of the Boston Colloquium for the Philosophy of Science 1966–1968*. 1969.
5. Robert S. Cohen and Marx W. Wartofsky (eds.), *Proceedings of the Boston Colloquium for the Philosophy of Science 1966–1968*. 1969.
6. Robert S. Cohen and Raymond J. Seeger (eds.), *Ernst Mach: Physicist and Philosopher*. 1970.
7. Milic Čapek, *Bergson and Modern Physics*. 1971.
8. Roger C. Buck and Robert S. Cohen (eds.), *PSA 1970. In Memory of Rudolf Carnap*. 1971.
9. A. A. Zinov'ev, *Foundations of the Logical Theory of Scientific Knowledge (Complex Logic)*. (Revised and enlarged English edition with an appendix by G. A. Smirnov, E. A. Sidorenka, A. M. Fedina, and L. A. Bobrova). 1973.
10. Ladislav Tondl, *Scientific Procedures*. 1973.
11. R. J. Seeger and Robert S. Cohen (eds.), *Philosophical Foundations of Science*. 1974.
12. Adolf Grünbaum, *Philosophical Problems of Space and Time*. (Second, enlarged edition). 1973.
13. Robert S. Cohen and Marx W. Wartofsky (eds.), *Logical and Epistemological Studies in Contemporary Physics*. 1973.
14. Robert S. Cohen and Marx W. Wartofsky (eds.), *Methodological and Historical Essays in the Natural and Social Sciences. Proceedings of the Boston Colloquium for the Philosophy of Science 1969–1972*. 1974.
15. Robert S. Cohen, J. J. Stachel, and Marx W. Wartofsky (eds.), *For Dirk Struik. Scientific, Historical and Political Essays in Honor of Dirk Struik*. 1974.
16. Norman Geschwind, *Selected Papers on Language and the Brain*. 1974
17. B. G. Kuznetsov, *Reason and Being: Studies in Classical Rationalism and Non-Classical Science*. 1987
18. Peter Mittelstaedt, *Philosophical Problems of Modern Physics*. 1976
19. Henry Mehlberg, *Time, Causality, and the Quantum Theory* (2 vols.). 1980.

R. Fawcett and Robert S. Cohen). 1984.
45. A. Ishimoto (ed.), *Japanese Studies in the History and Philosophy of Science.*
46. Peter L. Kapitza, *Experiment, Theory, Practice.* 1980.
47. Maria L. Dalla Chiara (ed.), *Italian Studies in the Philosophy of Science.* 1980.
48. Marx W. Wartofsky, *Models: Representation and the Scientific Understanding.* 1979.
49. Trân Duc Thao, *Phenomenology and Dialectical Materialism.* 1985.
50. Yehuda Fried and Joseph Agassi, *Paranoia: A Study in Diagnosis.* 1976.
51. Kurt H. Wolff, *Surrender and Catch: Experience and Inquiry Today.* 1976.
52. Karel Kosik, *Dialectics of the Concrete.* 1976.
53. Nelson Goodman, *The Structure of Appearance.* (Third edition). 1977.
54. Herbert A. Simon, *Models of Discovery and Other Topics in the Methods of Science.* 1977.
55. Morris Lazerowitz, *The Language of Philosophy. Freud and Wittgenstein.* 1977.
56. Thomas Nickles (ed.), *Scientific Discovery, Logic, and Rationality.* 1980.
57. Joseph Margolis, *Persons and Minds. The Prospects of Nonreductive Materialism.* 1977.
58. G. Radnitzky and G. Andersson (eds.), *Progress and Rationality in Science,* 1978.
59. Gerard Radnitzky and Gunnar Andersson (eds.), *The Structure and Development of Science.* 1979.
60. Thomas Nickles (ed.), *Scientific Discovery: Case Studies.* 1980.
61. Maurice A. Finocchiaro, *Galileo and the Art of Reasoning.* 1980.
62. William A. Wallace, *Prelude to Galileo.* 1981.
63. Friedrich Rapp, *Analytical Philosophy of Technology.* 1981.
64. Robert S. Cohen and Marx W. Wartofsky (eds.), *Hegel and the Sciences.* 1984.
65. Joseph Agassi, *Science and Society.* 1981.
66. Ladislav Tondl, *Problems of Semantics.* 1981.
67. Joseph Agassi and Robert S. Cohen (eds.), *Scientific Philosophy Today.* 1982.
68. Władysław Krajewski (ed.), *Polish Essays in the Philosophy of the Natural Sciences.* 1982.
69. James H. Fetzer, *Scientific Knowledge.* 1981.
70. Stephen Grossberg, *Studies of Mind and Brain.* 1982.
71. Robert S. Cohen and Marx W. Wartofsky (eds.), *Epistemology, Methodology, and the Social Sciences.* 1983.
72. Karel Berka, *Measurement.* 1983.
73. G. L. Pandit, *The Structure and Growth of Scientific Knowledge.* 1983.
74. A. A. Zinov'ev, *Logical Physics.* 1983.
75. Gilles-Gaston Granger, *Formal Thought and the Sciences of Man.* 1983.
76. R. S. Cohen and L. Laudan (eds.), *Physics, Philosophy and Psychoanalysis.* 1983.
77. G. Böhme et al., *Finalization in Science,* ed. by W. Schäfer. 1983.
78. D. Shapere, *Reason and the Search for Knowledge.* 1983.
79. G. Andersson, *Rationality in Science and Politics.* 1984.
80. P. T. Durbin and F. Rapp, *Philosophy and Technology.* 1984.
81. M. Marković, *Dialectical Theory of Meaning.* 1984.

114. A. Janik, *Style, Politics and the Future of Philosophy*. 1989.
115. F. Amrine (ed.), *Literature and Science as Modes of Expression*. 1989.
116. James Robert Brown and Jürgen Mittelstrass (eds.), *An Intimate Relation: Studies in the History and Philosophy of Science Presented to Robert E. Butts on his 60th Birthday*. 1989.
117. F. D'Agostino and I. C. Jarvie (eds.), *Freedom and Rationality: Essays in Honor of John Watkins*. 1989.
118. D. Zolo, *Reflective Epistemology: The Philosophical Legacy of Otto Neurath*. 1989.
119. Michael Kearn, Bernard S. Phillips and Robert S. Cohen (eds.), *George Simmel and Contemporary Sociology*. 1989.
120. Trevor H. Levere and William R. Shea (eds.), *Nature, Experiment, and the Sciences: Essays on Galileo and the History of Science in Honour of Stillman Drake*. 1989.